Astrochemistry
From the Big Bang to
the Present Day

Essential Textbooks in Chemistry   ISSN: 2059-7738

Orbitals: With Applications in Atomic Spectra
   by Charles Stuart McCaw

Principles of Nuclear Chemistry
   by Peter A C McPherson

Atmospheric Chemistry: From the Surface to the Stratosphere
   by Grant Ritchie

Astrochemistry: From the Big Bang to the Present Day
   by Claire Vallance

Forthcoming

Problems of Instrumental Analytical Chemistry: A Hands-On Guide
   by J M Andrade-Garda, A Carolsena-Zubieta, M P Gómez-Carracedo,
   M A Maestro-Saavedra, M C Prieto-Blanco and R M Soto-Ferreiro

Essential Textbooks in Chemistry

Astrochemistry
From the Big Bang to the Present Day

Claire Vallance
Oxford

World Scientific

Published by

World Scientific Publishing Europe Ltd.

57 Shelton Street, Covent Garden, London WC2H 9HE

Head office: 5 Toh Tuck Link, Singapore 596224

USA office: 27 Warren Street, Suite 401-402, Hackensack, NJ 07601

Library of Congress Cataloging-in-Publication Data
Names: Vallance, Claire.
Title: Astrochemistry : from the big bang to the present day / Claire Vallance, Oxford.
Description: New Jersey : World Scientific, 2016. | Series: Essential textbooks in chemistry | Includes bibliographical references and index.
Identifiers: LCCN 2016035656| ISBN 9781786340375 (hc : alk. paper) | ISBN 9781786340382 (pbk : alk. paper)
Subjects: LCSH: Cosmochemistry. | Astrophysics.
Classification: LCC QB450 .V35 2016 | DDC 523/.02--dc23
LC record available at https://lccn.loc.gov/2016035656

British Library Cataloguing-in-Publication Data
A catalogue record for this book is available from the British Library.

Copyright © 2017 by World Scientific Publishing Europe Ltd.

All rights reserved. This book, or parts thereof, may not be reproduced in any form or by any means, electronic or mechanical, including photocopying, recording or any information storage and retrieval system now known or to be invented, without written permission from the Publisher.

For photocopying of material in this volume, please pay a copying fee through the Copyright Clearance Center, Inc., 222 Rosewood Drive, Danvers, MA 01923, USA. In this case permission to photocopy is not required from the publisher.

Desk Editors: Suraj Kumar/Mary Simpson

Typeset by Stallion Press
Email: enquiries@stallionpress.com

Printed in Singapore by Mainland Press Pte Ltd.

Preface

This book, and its partner *Atmospheric Chemistry: From the Surface to the Stratosphere*, arose from a third year option course in Physical Chemistry entitled *Fundamentals of Astrochemistry and Atmospheric Chemistry* taught at the University of Oxford by myself and my colleague, Grant Ritchie, and from a short course on Astrochemistry that I taught at the University of Canterbury while on sabbatical leave in New Zealand in 2013. *Astrochemistry: From the Big Bang to the Present Day* aims to provide an accessible introduction to the rapidly growing field of astrochemistry. The book describes the chemical history of the Universe, our Solar System, and our planet, looks in some detail at the 'alien' chemistry occurring in interstellar gas clouds, and also details the experimental and theoretical laboratory-based methods that have allowed us to gain deeper insight into the chemistry occurring in these regions. *Atmospheric Chemistry: From the Surface to the Stratosphere* considers in detail the physics and chemistry of our contemporary planet, and in particular its atmosphere, explaining the chemistry and physics of the air that we breathe, that gives rise to our weather systems and climate, soaks up our pollutants and protects us from solar UV radiation.

Claire Vallance
March 2016

Author Biography

Claire Vallance is a Professor of Physical Chemistry in the Department of Chemistry at the University of Oxford, and Tutorial Fellow in Physical Chemistry at Hertford College. She holds B.Sc.(hons) and Ph.D. degrees from the University of Canterbury (Christchurch, NZ), where she worked on gas-phase molecular dynamics. She then moved to Oxford, where she held a Glasstone Research Fellowship and a Royal Society University Research Fellowship prior to her present appointment. Her current research interests include reaction dynamics, applications of velocity-map and spatial-map imaging to mass spectrometry, and the development of laser spectroscopy techniques employing optical microcavities for microfluidics and chemical sensing applications. Claire's Oxford tutorial teaching spans all of physical chemistry. She has given lecture courses on chemical kinetics, properties of gases, symmetry and group theory, reaction dynamics, and astrochemistry, as well as numerous outreach and public engagement lectures. She is author of over 90 journal articles, four book chapters, nine patents, an e-Textbook on *Symmetry and Group Theory*, and recently co-edited the textbook *Tutorials in Molecular Reaction Dynamics*.

Acknowledgements

I would very much like to thank two of my students, Jakub Sowa and David Heathcote, for rising well above and beyond the call of duty to proof read this book in draft, and for the many helpful suggestions they both made for improvements. I would also like to thank Grant Ritchie, for talking me into writing the book in the first place, and for his unstinting enthusiasm as the project has proceeded. Laurent Chaminade, Catharina Weijman, Suraj Kumar, and Mary Simpson from World Scientific Publishing deserve considerable gratitude for their unending patience as agreed deadlines came and agreed deadlines went... I hope you feel the end result is worth it. My research group and my partner, Stephen Clarke, deserve many thanks for putting up with the increased level of general grumbling that goes along with trying to write a book at the same time as having a very full-time job and life. I am grateful to Nick Green, Oxford Chemistry's Director of Studies, for allowing me to adapt tutorial and exam questions I have developed previously for the Oxford Chemistry course for use in some of the book's end-of-chapter problems. Finally, both Grant and I would like to thank the undergraduate students at the Universities of Oxford and Canterbury who have attended our various lecture courses, asked questions that made us think, and provided feedback that enabled us to improve both our lecture courses and these books.

Contents

Preface		v
Author Biography		vii
Acknowledgements		ix
List of Figures		xvii
List of Tables		xxiii

1. Measuring the Universe ... 1
   - 1.1 Studying the Universe via spectroscopy ... 1
     - 1.1.1 Line positions ... 1
     - 1.1.2 Line intensities ... 3
     - 1.1.3 Principle of operation of an astronomical spectrograph ... 4
     - 1.1.4 Spectral windows for Earth-based observations ... 4
   - 1.2 Döppler shift ... 5
   - 1.3 Döppler lineshape ... 6
   - 1.4 Döppler shift, the Hubble constant, and the age of the Universe ... 7
   - 1.5 Questions ... 8
     - 1.5.1 Essay-style questions ... 8
     - 1.5.2 Problems ... 9

2. From the Big Bang to the First Atoms ... 11
   - 2.1 The very early Universe: The building blocks of matter ... 11
   - 2.2 The nature of the expanding Universe ... 11
   - 2.3 The first particles ... 13

	2.4	Hydrogen and helium nuclei		15
	2.5	The first atoms		15
	2.6	Questions		16
		2.6.1	Essay-style questions	16
		2.6.2	Problems	16
3.	Stars and the Creation of the Higher Elements			19
	3.1	Star formation and the nucleosynthesis of heavier elements		19
	3.2	Dispersion of the chemical elements into interstellar space		23
	3.3	Cosmic abundance of the elements		25
	3.4	Questions		26
		3.4.1	Essay-style questions	26
		3.4.2	Problems	26
4.	Interstellar Chemistry — Molecules in Space			29
	4.1	The interstellar medium		29
		4.1.1	Diffuse interstellar medium	29
		4.1.2	Giant molecular clouds	30
		4.1.3	Circumstellar medium	32
	4.2	Chemistry in interstellar space		32
	4.3	Molecular synthesis in interstellar gas clouds		34
	4.4	Ionisation processes in the interstellar medium		34
	4.5	Gas-phase chemical reactions in the interstellar medium		36
	4.6	Bond-forming reactions		36
		4.6.1	Radiative association	36
		4.6.2	Associative detachment	37
		4.6.3	Dust-grain-catalysed reactions	37
	4.7	Bond breaking reactions		38
		4.7.1	Photodissociation and collisional dissociation	38
		4.7.2	Dissociative recombination	39
	4.8	Rearrangement reactions		39
		4.8.1	Charge transfer	39
		4.8.2	Neutral reactions	40
		4.8.3	Ion–molecule reactions	40

			4.8.3.1	Hydrogen atom abstraction	40
			4.8.3.2	Proton transfer	41
			4.8.3.3	Carbon insertion	41
			4.8.3.4	Rearrangement reactions	42
	4.9	Neutralisation processes in the interstellar medium			42
	4.10	Summary			43
	4.11	Questions			43
		4.11.1	Essay-style questions		43
		4.11.2	Problems		44
5.	Laboratory-Based Astrochemistry: Theory				47
	5.1	Laboratory-based astrochemistry			47
	5.2	The grand challenge: Chemical modelling of giant molecular clouds			48
		5.2.1	The search for biological molecules		49
		5.2.2	The diffuse interstellar bands (DIBs)		50
	5.3	Theoretical astrochemistry I: Spectroscopic data			51
		5.3.1	Rotational transition frequencies		53
		5.3.2	Vibrational transition frequencies		55
		5.3.3	Electronic transition frequencies		58
		5.3.4	Transition intensities		58
	5.4	Theoretical astrochemistry II: Kinetic and dynamical data			59
		5.4.1	Types of collision		60
		5.4.2	Relative velocity		60
		5.4.3	Collision energy, total kinetic energy, and conservation of linear momentum		61
		5.4.4	Conservation of energy and energy available to the products		62
		5.4.5	Impact parameter, b, and opacity function, $P(b)$		62
		5.4.6	Collision cross-section, σ_c		63
		5.4.7	Reaction cross-section, σ_r		64
		5.4.8	The excitation function, $\sigma_r(E_{\text{coll}})$, and the thermal rate constant, $k(T)$		65
			5.4.8.1	Exoergic with no barrier	66
			5.4.8.2	Endoergic or exoergic with a barrier	67

	5.4.9	Orbital angular momentum, **L**, and conservation of angular momentum		67
	5.4.10	The interaction potential and its effect on the collision cross-section		69
	5.4.11	Atomic and molecular interactions		71
	5.4.12	The potential energy surface for a polyatomic system		71
	5.4.13	Construction of the potential energy surface		73
	5.4.14	The potential energy surface and the collision dynamics		73
	5.4.15	The potential energy surface for a linear triatomic system		75
	5.4.16	Reactive and non-reactive trajectories across the potential energy surface		77
	5.4.17	General features of potential energy surfaces		79
	5.4.18	Examples of potential energy surfaces for real chemical systems		80
		5.4.18.1	The simplest chemical reaction: $H + H_2 \to H_2 + H$	80
		5.4.18.2	Photodissociation of NO_2	81
		5.4.18.3	$H + CO_2 \to OH + CO$ and $OH + SO \to H + SO_2$	82
		5.4.18.4	The $Ar + H_2^+ \to ArH^+ + H$ reaction	82
	5.4.19	Orbital angular momentum, centrifugal barriers and the effective potential		84
	5.4.20	A simple model for the rate of ion–molecule reactions		85
	5.4.21	Reaction cross-sections from quasi-classical trajectory calculations		89
5.5	Summary			89
5.6	Questions			90
	5.6.1	Essay-style questions		90
	5.6.2	Problems		90
6.	**Laboratory-Based Astrochemistry: Experiment**			**97**
6.1	Experimental astrochemistry I: Spectroscopic data			97
	6.1.1	Molecular beams		98
		6.1.1.1	Effusive sources	99
		6.1.1.2	Supersonic sources	100

		6.1.2	Fourier-transform microwave spectroscopy	102

		6.1.2	Fourier-transform microwave spectroscopy	102
		6.1.3	Laser-induced fluorescence	103
		6.1.4	Resonance-enhanced multiphoton ionization (REMPI)	104
		6.1.5	Cavity-enhanced absorption spectroscopy methods	106
			6.1.5.1 Cavity ring-down spectroscopy	107
			6.1.5.2 Cavity-enhanced absorption spectroscopy	109
		6.1.6	Molecular size considerations	110
	6.2	Experimental astrochemistry II: Gas-phase kinetic and dynamical data		111
		6.2.1	Ion cyclotron resonance mass spectrometry	112
			6.2.1.1 The ion cyclotron resonance technique	112
			6.2.1.2 Measuring ion–molecule rate constants via ICR-MS	115
		6.2.2	The flowing afterglow technique	116
		6.2.3	The selected-ion flow tube	117
		6.2.4	The CRESU method	117
		6.2.5	Coulomb crystals	118
		6.2.6	Neutral reactions	120
	6.3	Experimental astrochemistry III: Dust-grain chemistry		121
		6.3.1	Ice structures via infrared spectroscopy	121
		6.3.2	Thermodynamics of adsorption and desorption via temperature-programmed desorption	123
		6.3.3	Photoinitiated molecular synthesis in interstellar ice analogues	124
		6.3.4	Formation of H_2 on ice surfaces	124
	6.4	Case study: Ethylene glycol		125
	6.5	Summary		129
	6.6	Questions		129
		6.6.1	Essay-style questions	129
		6.6.2	Problems	130
7.	Formation of the Solar System and the Evolution of Earth			135
	7.1	The Solar nebula		136
	7.2	The protoplanetary disk		138

7.3	Formation of the planets		138
7.4	The early Earth, and formation of the Moon		143
	7.4.1	The Moon's orbit and tidal locking	145
7.5	The layered structure of the Earth		147
	7.5.1	The core and the Earth's magnetic field	148
	7.5.2	The mantle	149
	7.5.3	The crust	152
		7.5.3.1 Divergent plate boundaries	153
		7.5.3.2 Convergent plate boundaries	153
	7.5.4	The primordial atmosphere	155
7.6	Oceans and tides		155
7.7	Erosion and weathering		158
7.8	Life and the oxygen atmosphere		160
7.9	Fossilisation and fossil fuels		164
7.10	Other Solar systems		166
7.11	Further reading		169
7.12	Questions		169
	7.12.1	Essay-style questions	169
	7.12.2	Problems	170

Appendix A	Rates of Chemical Reactions	175
A.1	Reactions occurring in a single step	175
A.2	Reactions occurring in multiple steps	177
A.3	Experimental kinetics studies	179

Appendix B	The Variation Principle and the Linear Variation Method	181
B.1	The variation principle	181
B.2	The linear variation method	182

Appendix C	Mass-Weighted Coordinates and the Skew Angle	185

Answers to Numerical Problems   189

Index   195

List of Figures

1.1 (a) High-resolution spectrum of the Sun, recorded at the National Solar Observatory, Kitt Peak, USA; (b) schematic diagram of an astronomical spectrograph; (c) absorption spectrum of the Earth's atmosphere.   2

1.2 Determining the speed and sense of Jupiter's rotation through a measurement of the Döppler shift of the hydrogen Lyman α line.   6

1.3 Effect of temperature on the Döppler lineshape of a sample of gas. The arrows show the direction of propagation of the excitation light through a sample of cold gas (left), and hot gas (right). The resulting linewidths are shown below in each case.   7

1.4 A Hubble plot of data from a series of Type IA supernovae.   8

2.1 In the expansion of the Universe, matter does not expand into an existing space, as shown in (a), but instead space itself expands at all points simultaneously, as shown in (b).   12

3.1 (a) Star-forming region in the Small Magellanic Cloud, 210,000 light years from Earth (NASA image SPD-HUBBLE-STScI-2005-35a); (b) Star-forming region associated with the Rosette Nebula, a stellar nursery in the Monoceros constellation, 5000 light years from Earth (NASA image 450093main_hobys_rosette_05-A3_3d); (c) Hubble ultra-deep field image showing a spectacular array of galaxies within an area of sky 3.4 arc minutes in diameter (created for NASA by the Space Telescope Science Institute under Contract NAS5-26555).   20

3.2 Binding energy per nucleon for the chemical elements.   22

3.3 Life cycles of typical low-mass and high-mass stars.   23

3.4 NASA images of low-mass and high-mass stars at various points in their life cycles: (a) Mira, a red giant; (b) NGC6751 white dwarf and planetary nebula; (c) M57 ring nebula (and white dwarf); (d) the shadow cast by a black dwarf; (e) Betelgeuse, a red supergiant; (f) and (g) Cassiopeia A, a remnant of a supernova 11,000 years ago — the neutron star is visible in the centre.   24

3.5 The cosmic abundance of the elements.   25

4.1 Giant molecular clouds: (a) a region of the M17 molecular cloud within the Omega nebula of Sagittarius. The Sagittarius constellation is around 5500 light years away from Earth, and the image spans around 3 light years. Green, red, and blue hues correspond to abundant hydrogen and trace amounts of sulphur and oxygen, respectively; (b) the Barnard 68 molecular cloud. The cloud appears dark against the background of the stars, as visible light is unable to penetrate through the cloud due to absorption and scattering by hydrogen and dust grains (NASA image gallery).   30

4.2 Spectrum of interstellar dust from the W33A dust-embedded massive young star.   38

4.3 Examples of polycyclic aromatic hydrocarbons.   42

5.1 Kinetic model of H_2O chemistry in star-forming environments.   50

5.2 The sequence of steps involved in calculating a rotational absorption or emission spectrum.   53

5.3 The vibrational potential energy curve of a diatomic molecule is a Morse potential, but is often approximated by a harmonic oscillator potential.   55

5.4 The impact parameter, b, denotes the smallest perpendicular distance between the reactants in the absence of interactions.   63

5.5 The collision cross-section defines the cross-sectional area within which the centres of two particles must lie if they are to collide.   64

5.6 Volume elements involved in the definition of the reaction cross-section.   65

5.7 Generic excitation functions for exoergic reactions (a) without an activation barrier and (b) with a barrier; (c) At low relative velocities, attractive forces act over relatively long periods of time to pull particles together, leading to a large reaction cross-section; at high relative velocities, there is little time for

List of Figures

	attractive interactions to act, and trajectories are deflected only slightly, if at all, leading to small or 'hard-sphere' reaction cross-sections.	66
5.8	The line of centres of two approaching particles rotates about their centre of mass, giving rise to orbital angular momentum.	68
5.9	From left to right: the interaction potential, an example trajectory, and the velocity-dependent collision cross-section for a hard-sphere collision.	69
5.10	The interaction potential, an example trajectory, and the velocity-dependent collision cross-section for a collision governed by a Lennard-Jones potential.	70
5.11	(a) Coordinates needed to define the relative positions of two, three, and four atoms; (b) schematic illustration of a potential energy curve for a diatomic collision, and a potential energy surface for a many-body collision.	72
5.12	(a) Potential energy curve plotted as a function of the r_{AB} distance for the entrance channel of the reaction A + BC; (b) schematic PES for a reaction A + BC → AB + C, shown as a contour plot. The atomic positions at various points on the surface are shown, and the transition state (TS) is marked; (c) The reaction coordinate and potential energy profile for a reaction A + BC → AB + C.	76
5.13	Four different trajectories across the potential energy surface for an A + BC reaction.	77
5.14	Schematic of a general potential energy surface, with the various possible topological figures labelled.	79
5.15	(a) Potential energy surface for the reaction H + H$_2$ → H$_2$ + H, shown as both a contour plot and three-dimensional plot. The reaction coordinate is shown in blue, and dividing surfaces for reactants, transition state, and products are shown in purple; (b) potential energy surface and dissociation mechanism for the photodissociation of NO$_2$; potential energy surfaces for the reactions (c) H + CO$_2$ → OH + CO (contour plot) and (d) H + SO$_2$ (three-dimensional surface plot).	81
5.16	Potential energy surface for the reaction Ar + H$_2^+$ → ArH$^+$ + H at various different Ar–H–H angles.	83
5.17	The effective potential for low and high values of the impact parameter, b, and therefore the orbital angular momentum, L,	

showing the appearance of a centrifugal barrier at large values of the orbital angular momentum. 85

5.18 (a) Since the height of the centrifugal barrier depends on the initial impact parameter, b, there will be a maximum impact parameter $b_{\max}$ beyond which the reactants do not have sufficient kinetic energy to overcome the barrier and react; (b) the Langevin reaction cross-section as a function of reactant relative velocity v_{rel}. 86

6.1 Reducing the temperature of a sample reduces the number of occupied rotational levels, often yielding a dramatic simplification in the rotational spectrum. 98

6.2 Preparation of effusive and supersonic molecular beams, and the resulting molecular speed distributions. 99

6.3 (a) Shock-wave structure of a supersonic molecular beam; (b) an electroformed nickel skimmer; (c) use of a skimmer to deflect supersonic shock waves and to collimate the molecular beam. 102

6.4 (a) Schematic of a Fourier-transform microwave spectrometer — see text for details; (b) the microwave excitation pulse excites the beam molecules coherently into a variety of rotational states with a well-defined phase relationship between their rotational motion; (c) the excited molecules emit radiation coherently to return to the ground state, resulting in an interference pattern or free induction decay in the time domain, which can be Fourier transformed into the frequency domain to yield the rotational spectrum. 103

6.5 (a) In LIF, molecules undergo laser excitation to a fluorescent state, and the resulting fluorescence intensity is used to probe the original population of the lower-state energy levels; (b) LIF spectrum of $OH(v = 2)$ reaction products formed in the reaction $O(^1D) + H_2 \rightarrow OH + H$. 104

6.6 (a) Schematic of a (2 + 1)REMPI transition, in which two photons are used to access an intermediate state, before a third photon ionises the molecule; (b) REMPI-TOF signal for jet-cooled HF. 105

6.7 (a) Schematic of a conventional absorption spectroscopy measurement; (b) schematic of the experimental setup for a cavity ring-down or cavity-enhanced absorption measurement. 106

6.8 (a) Schematic of an ion cyclotron resonance mass spectrometer, showing the excitation, detection, and trapping plates; (b) principle of detection within an ICR mass spectrometer. 113
6.9 Damping of the ICR signal with time due to reactive ion–molecule collisions within the ICR cell. 115
6.10 Schematic of (a) a flowing afterglow instrument and (b) a selected-ion flow tube (SIFT). 116
6.11 Schematic of a CRESU instrument. 118
6.12 (a) A closed optical loop is used for laser cooling of atoms. (b) A magneto-optical trap (MOT) is used for Döppler cooling and trapping of cooled atoms. (c) A Coulomb crystal of Rb atoms stored within a MOT. 119
6.13 (a) Transmission-mode and (b) reflection-mode IR measurements on laboratory ice samples; (c) the peak in a TPD measurement results from competition between the increase in desorption rate constant with temperature and the decrease in the number of adsorbed molecules; (d) sample set of data for temperature-programmed desorption of water ices from an alumina surface, recorded as part of a study by Tzvetkov *et al.* on the interaction of glycine with ice nanolayers. 122
6.14 Lowest-energy conformation of ethylene glycol. 125
6.15 (a) Predicted energy levels and transitions for ethylene glycol, taken from Christen *et al.*; (b) a section of the interstellar ethylene glycol spectrum recorded by Hollis *et al.* 127
7.1 Ball and spring analogy of angular momentum transfer in an accretion disk. The spring represents an attractive force (e.g. a magnetic force in some models). As the inner particle orbits faster than the outer particle, the spring is stretched. The restoring force slows down the inner particle and speeds up the outer particle, at the same time causing the inner particle to move to smaller orbital radius and the outer particle to move to larger orbital radius. 137
7.2 The protoplanetary disk surrounding the young star HL Tauri in the constellation Taurus, 450 light years from Earth. The dark rings most probably correspond to the orbits of newly forming planets as they accrete smaller bodies in their path. Credit: Atacama Large Millimeter/Submillimeter Array (ALMA) (European Southern Observatory (ESO)/National

	Astronomical Observatory of Japan (NAOJ)/National Radio Astronomy Observatory (NRAO)).	141
7.3	The Earth's 23.5° tilt gives rise to the seasons with the Northern and Southern hemispheres being exposed to differing amounts of sunlight as the Earth orbits the Sun.	145
7.4	(a) Tidal bulges are caused by the gravitational attraction between a planet and its Moon. In a tidally locked system, both planet and moon rotate at the same frequency as they orbit, and the tidal forces between the bulges act along their line of centres. (b) If the planet rotates faster than it orbits (currently the case for the Earth–Moon system), then while gravity acts primarily along the planet–moon axis, the 'additional' tidal forces between the bulges are off axis.	146
7.5	The layered structure of the Earth.	148
7.6	(a) Convection currents in the Earth's core give rise to a dipolar magnetic field, currently aligned at a small angle to the Earth's rotation axis; (b) the Earth's magnetic field deflects the solar wind, protecting the Earth's atmosphere and surface. The solar wind is deflected at an imaginary surface known as the bow shock, and the interaction between the solar wind and the magnetic field alters the shape of the dipolar field as shown.	150
7.7	(a) Mantle minerals: Peridotite consists primarily of olivine and pyroxene minerals; (b) phase diagram illustrating the phase transitions from olivine to wadsleyite and ringwoodite that occur deep in the Earth's mantle; (c) crust minerals: The Earth's crust consists primarily of granite and basalt rock.	151
7.8	(a) A phospholipid consists of a polar, hydrophilic 'head group' and a non-polar, hydrophobic 'tail group'; (b) these relatively simple building blocks can self-assemble into a variety of complex structures in solution, depending on the shape and size of their head and tail groups.	161
C.1	Jacobi coordinates: r is the BC distance; R is the distance from A to the centre of mass of BC; and φ is the angle between r and R.	186
C.2	An A + BC potential energy surface plotted in mass-weighted coordinates, in which the r_{AB} and r_{BC} axes are plotted at a skew angle β.	187

List of Tables

4.1	Molecular species identified in giant molecular clouds. Unconfirmed detections are denoted with a question mark.	31
5.1	Values of a and n for molecular interactions with potential energy $V(r) = -a/r^n$.	71
6.1	Table of data for ethylene glycol.	127
7.1	Relative abundances and standard melting points of atomic and molecular species in a dense interstellar cloud, as calculated by Millar and coworkers.	139
7.2	The elemental composition of the Earth.	143
7.3	Approximate average composition of granite and basalt rocks in the Earth's crust.	154
7.4	Major components of the modern atmosphere of Earth.	164

Chapter 1

Measuring the Universe

1.1 Studying the Universe via spectroscopy

In contrast to many other areas of chemistry and physics, we cannot carry out any active field-based experiments to study the chemistry of the Universe. Instead, all of the information we have comes from passive observations. The data from telescopes arrives in the form of spectroscopic signatures recorded in various portions of the electromagnetic spectrum (ultraviolet (UV), visible, infrared, microwave, etc.) for different regions of space (stars, interstellar regions, and so on), and can be interpreted in order to establish the atomic and molecular composition of these regions. As an example, Figure 1.1(a) shows a high-resolution spectrum of the Sun, recorded at the National Solar Observatory at Kitt Peak, USA. Each element present in the Sun gives rise to a unique set of emission lines, and by analysing the positions and intensities of the lines observed in the spectrum, it is possible to determine the elemental composition of the Sun.

1.1.1 Line positions

A given atom or molecule absorbs and emits light at a characteristic set of frequencies, yielding a unique 'fingerprint' that can be used to detect its presence in a given region of space. Of course, in order to identify a molecule in space, its characteristic absorption/emission frequencies must be known. Sometimes, the line positions can be determined by recording the spectra of the species of interest in an Earth-based laboratory — we will look at a number of suitable methods for making such measurements

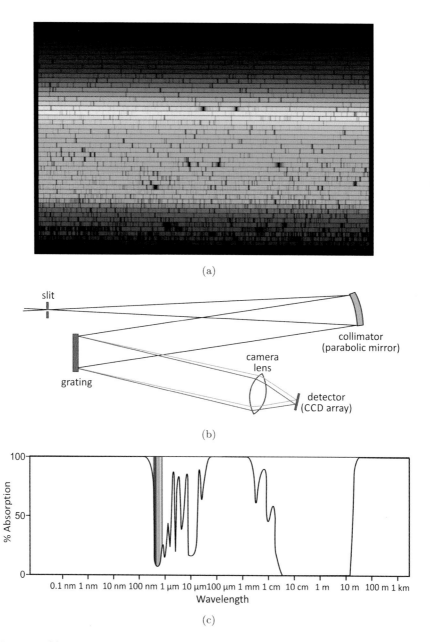

Fig. 1.1 (a) High-resolution spectrum of the Sun, recorded at the National Solar Observatory, Kitt Peak, USA; (b) schematic diagram of an astronomical spectrograph; (c) absorption spectrum of the Earth's atmosphere.

in Chapter 6. However, many of the chemical species formed in space are extremely unstable at terrestrial temperatures and pressures, and cannot be formed in sufficient quantities for the spectra to be recorded. In these cases, we must rely on theoretical predictions of the line positions. We will discuss the methods for calculating atomic and molecular spectra in Chapter 5.

1.1.2 Line intensities

The spectrum of the Sun shown in Figure 1.1(a) is an emission spectrum, consisting of bright emission lines against a dark background. Given the relevant (measured or calculated) emission cross-sections or lifetimes for the emitting states, the line intensities provide information about the number densities of the various electronically excited atoms present in the Sun. If we assume that the atoms are in thermal equilibrium at a given temperature T, then the Boltzmann distribution (discussed later in Section 5.3.4) can be used to determine the total density of each type of atom, regardless of their state of excitation.

Emission spectra can be recorded from any region of space in which the temperature is high enough for there to be an appreciable density of electronically excited atoms, or electronically, vibrationally, or rotationally excited molecules. Colder regions of space may be probed by studying the amount of starlight absorbed as it passes through the region of interest. In this case, the presence of atoms or molecules that absorb light will lead to characteristic dark lines against the bright background of transmitted starlight. The line intensities may be related to the number density of the absorbing species via the Beer–Lambert law, which describes the fraction of light transmitted through a sample of gas at a particular wavelength λ.

$$\frac{I(\lambda)}{I_0(\lambda)} = \exp(-n\sigma l) \tag{1.1}$$

where $I(\lambda)$ and $I_0(\lambda)$ are the transmitted and incident intensities of light at wavelength λ, $n = N/V$ is the number density (number of molecules, N, per unit volume, V) of the molecule of interest, σ is its absorption cross-section at the wavelength of interest, and l is the path length of the light through the sample. The appropriate path length to use will often be known from other measurements on the astronomical body of interest. In some cases values for σ will be known from terrestrial measurements, in which case the only remaining unknown is the number density of the species of interest, which may therefore be determined from the measured spectrum. In most cases, extensive fitting of astronomical spectrum is required in order to extract the best-fit values for the various parameters of interest.

1.1.3 Principle of operation of an astronomical spectrograph

While there are many designs of astronomical spectrograph, they all share the same basic principles of operation.[1] As shown in Figure 1.1(b), the essential elements are: a slit; a collimator, usually a parabolic mirror; a dispersing element, usually a reflection grating; and a camera comprising a lens or lenses that focus the spectrum onto a position-sensitive detector. The image of the astronomical object of interest is focused onto the slit, and the diverging beam of light from the slit continues onto the collimator. The instrument is designed so that the focal length and diameter of the collimator are such that the light from the slit fills the collimator. The collimated (parallel) beam continues to the grating (or other dispersing element), where it is dispersed into its constituent wavelengths. The dispersed light is focused onto the position-sensitive detector, usually a sensitive charge-coupled device (CCD) array, from which the signal can be read out to a computer, yielding the spectrum of the astronomical object. The spectral resolution is determined by the slit width and the grating resolution, and may be optimised for a particular application, and gratings may be designed for optimal operation over particular wavelength ranges. Most modern spectrographs are somewhat more complex than the simple prototype shown in Figure 1.1(b). For example, the slit may be replaced by an optical fibre or fibre bundle, filters may be included in order to remove undesired wavelengths, and so on, but the basic operating principles are similar.

1.1.4 Spectral windows for Earth-based observations

The transmission spectrum of the Earth's atmosphere is shown in Figure 1.1(c). Ground-based telescopes are limited to studying regions of the electromagnetic spectrum in which the Earth's atmosphere does not

---

[1] P. Massey and M. M. Hanson, Astronomical spectroscopy, in *Planets, Stars, and Stellar Systems* (Springer, 2011).

absorb. These include the UV–visible window spanning 300–900 nm, corresponding to electronic transitions; two infrared windows from 1 to 5 μm and 8 to 20 μm which can be used to probe vibrational transitions; a window in the microwave and millimetre-wave region from 1.3 to 0.35 mm, which can be used to probe rotational transitions; and a radio-wave window from 2 to 10 m which contains information on transitions between atomic hyperfine levels, such as the 21 cm line in atomic hydrogen. Space telescopes are not subject to such restrictions, and can observe in any region of the spectrum. In general, the microwave region is the most information rich, and therefore the most useful for studying small molecules in space, followed by the infrared and UV–visible regions.

Atoms and molecules may be identified through their absorption or emission frequencies and quantified by their absorption or emission intensities. There are one or two additional features of analysing astronomical spectral data which differ from those encountered in more everyday terrestrial spectroscopy. Firstly, most transitions will be Döppler shifted, and secondly, the lineshape can provide a great deal of information on the motion of the body under study.

1.2 Döppler shift

For velocities significantly less than the speed of light,[2] the observed wavelength of light absorbed or emitted by a moving object is Döppler shifted by an amount

$$\frac{\delta\lambda}{\lambda} = \frac{v_{\text{source}}}{c} \tag{1.2}$$

where λ is the transition wavelength, $\delta\lambda$ is the Döppler shift, v_{source} is the velocity of the source relative to the observer, and c is the speed of light. The Döppler shift must be taken into account when identifying spectroscopic lines from telescope data. Since the Universe is expanding, such that all other astronomical objects are moving away from the Earth, often the lines will appear at wavelengths that are significantly red-shifted from those expected for a stationary sample. If the velocity of the source is known then the wavelengths can be corrected for comparison with lab-based or calculated spectroscopic data. If the velocity of the source is not known then

---

[2] For velocities greater than around $0.7c$, a relativistic correction is required.

Fig. 1.2 Determining the speed and sense of Jupiter's rotation through a measurement of the Döppler shift of the hydrogen Lyman α line.

lines must be identified through a process of 'pattern matching', and when a match is found they may be used to determine the velocity of the source.

A good example of velocity measurements made via the observed Döppler shift of a spectral line is the determination of the rotational speed of Jupiter (see Figure 1.2). During the rotation, one 'side' of the planet (as observed from Earth) is moving towards the Earth and will be blue-shifted, while the other side is moving away from the Earth and will be red-shifted. By comparing the Döppler shift of the hydrogen Lyman α line from each side of the planet, we can determine both the speed of rotation and its direction. Data from this measurement is included in Problem P1.2 at the end of the chapter.

1.3 Döppler lineshape

Within a sample there will usually be a distribution of velocities, meaning that each molecule will have a slightly different Döppler shift. As illustrated in Figure 1.3, this leads to line broadening over and above the natural linewidth. For a sample of molecules of mass m with a Maxwell–Boltzmann velocity distribution at temperature T, the Döppler linewidth is given by

$$\delta\lambda = \frac{2\lambda}{c}\left(\frac{2k_\mathrm{B}T\ln 2}{m}\right)^{1/2} \qquad (1.3)$$

In principle, therefore, the Döppler linewidth may be used to determine the temperature of a sample. In practice, the situation is often complicated by local turbulence. Turbulence also leads to a distribution of molecular velocities, and is often the dominant factor in determining the Döppler linewidth.

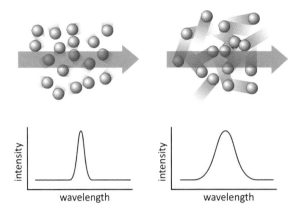

Fig. 1.3 Effect of temperature on the Döppler lineshape of a sample of gas. The arrows show the direction of propagation of the excitation light through a sample of cold gas (left), and hot gas (right). The resulting linewidths are shown below in each case.

1.4 Döppler shift, the Hubble constant, and the age of the Universe

Hubble measured the Döppler red-shift for a number of galaxies, and found that when red-shift was plotted against distance for galaxies spanning a range of distances from Earth, the plot yielded a straight line. Hubble plots have since been constructed for a variety of astronomical objects. A Hubble plot of a series of Type IA supernovae is shown in Figure 1.4.

The result is summarised in the Hubble law:

$$v = Hd \qquad (1.4)$$

Here, v is the radial velocity, H is the Hubble constant, and d is the distance from the observer. The equation summarises the rate of recession of other galaxies, and therefore the rate at which the Universe is expanding. The most recent determination of the Hubble constant yields a value of 23 km s^{-1} Mly^{-1}, with an error of around 10%. Other estimates give a value between 19 and 21 km s^{-1} Mly^{-1}. If we assume that the rate of expansion of the Universe has remained constant (this turns out not to be an entirely correct assumption, as we shall see later), then the age of the Universe is simply the inverse of the Hubble constant, giving an age of 13.8 billion years.

Over the next few chapters we will explore the evolution of the Universe, and particularly our own planet, from these origins 13.8 billion years ago. We will begin our astrochemical voyage in Chapters 2 and 3 by investigating

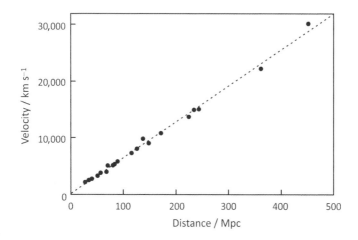

Fig. 1.4 A Hubble plot of data from a series of Type IA supernovae.
Source: Data from 'Using Type IA supernova light curve shapes to measure the Hubble constant', A. G. Riess, W. H. Press, and R. P. Kirshner, *Astrophys. J.*, **438**, L17 (1995).

the very early stages of the Universe and the events leading to synthesis of the chemical elements, before taking a tour in Chapter 4 of the interstellar medium, home to a rich chemistry of alien molecules that could not possibly survive on Earth. The conditions of ultra-low temperature and pressure in the interstellar medium are difficult to reproduce on Earth, and pose considerable challenges for studying the chemistry in such regions. Many ingenious approaches have been developed, and we will look in some detail in Chapters 5 and 6 at the various theoretical and experimental methods that have been applied to studying the many different chemical processes that occur in interstellar space. In Chapter 7, we will consider the formation of the Solar System, and in particular the formation and evolution of Earth, from its early stages as a molten protoplanet through to our contemporary home, complete with oceans, mountains, an oxygen-based atmosphere, and a rich diversity of life.

1.5 Questions

1.5.1 Essay-style questions

Q1.1 Explain the types of information on different regions of space that can be obtained from spectroscopic measurements using telescopes.

Discuss the relative merits of ground and space-based telescopes as part of your answer.

Q1.2 Explain the term *Döppler shift* in the context of spectroscopic measurements of astronomical objects. What information does the Döppler shift provide? Include in your answer a discussion of Döppler lineshapes.

1.5.2 Problems

P1.1 *Atomic transitions and atmospheric windows*

(a) Use the Rydberg equation, $\Delta E = hcR(1/n_1^2 - 1/n_2^2)$, where h is Planck's constant, c is the speed of light, R is the Rydberg constant, and n_1 and n_2 are the quantum numbers of the lower and upper states in the electronic transition, to calculate the energies ΔE_{12} of the first transition in the Lyman, Balmer, and Paschen series in the emission spectrum of atomic hydrogen, which terminate on $n = 1, 2,$ and 3, respectively.

(b) Which of these lines can be observed from an Earth-based telescope?

P1.2 *Döppler shift and the rotational speed of Jupiter*

The Hα transition is the transition from the $n = 3$ to $n = 2$ level in atomic hydrogen. The transition has been used to determine the rotation speed of Jupiter.

(a) Use the Rydberg equation to calculate the wavelength of the Hα transition to three decimal places.

(b) The rotation speed of a point on the surface of Jupiter at the equator is $43{,}000 \, \text{km h}^{-1}$. Calculate the Döppler shifts for the Hα line observed from points on the extreme left and extreme right of Jupiter's equator, as viewed from Earth. What would be the Döppler shift for a point observed in the centre of the planet as viewed from Earth?

(*Note*: ignore the contribution to the Döppler shift from translational motion of Jupiter relative to Earth).

P1.3 *Temperatures from Döppler widths of spectral lines*

The microwave spectrum of an interstellar gas cloud has been recorded in the region containing CO transitions. The width of the $J = 1 \to J = 0$ line, appearing at a wavelength of 2.6007576 mm, has been measured to be 2.7262157 nm. Determine the temperature of the molecular cloud. The mass of CO is 28.0101 g mol^{-1}.

P1.4 *Temperature and energy level populations*

The star Rigel has a surface temperature of 11,000 K. Calculate the relative populations of the two levels involved in the Hα transition (see Problem P1.2) at this temperature.

(*Note*: The ratio of populations, n_i/n_j of two states with energies ε_i and ε_j and degeneracies g_i and g_j at a temperature T is given by the Boltzmann distribution

$$\frac{n_i}{n_j} = \frac{g_i}{g_j} \exp\left(\frac{-(\varepsilon_i - \varepsilon_j)}{k_B T}\right)$$

where k_B is Boltzmann's constant).

P1.5 *Determination of the Hubble constant*

The table below reproduces some of the data from Hubble's original 1929 paper entitled 'A relation between distance and radial velocity among extra-galactic nebulae' (PNAS 1929, pp. 168–173). The table lists distances (in parsecs, where 1 parsec = 3262 light years) and velocities (in km s^{-1}) for a number of extra-galactic nebulae.

Distance (10^6 parsecs)	Velocity (km s^{-1})	Distance (10^6 parsecs)	Velocity (km s^{-1})
0.032	+170	0.9	+650
0.034	+290	0.9	+150
0.214	−130	0.9	+500
0.263	−70	1.0	+920
0.275	−185	1.1	+450
0.275	−220	1.1	+500
0.45	+200	1.4	+500
0.5	+290	1.7	+960
0.5	+270	2.0	+500
0.63	+200	2.0	+850
0.8	+300	2.0	+800
0.9	−30	2.0	+1090

Use the data to construct a Hubble plot and to determine the Hubble constant, and hence the approximate age of the Universe.

Chapter 2

From the Big Bang to the First Atoms

2.1 The very early Universe: The building blocks of matter

The story of planet Earth begins with the creation of the Universe during the Big Bang. Immediately after the Big Bang, during a brief period of time known as the *inflationary epoch*, the Universe underwent extremely rapid exponential expansion, increasing its volume by a factor of at least 10^{78} and expanding from subatomic dimensions to around the size of a grapefruit. The inflationary epoch lasted until around 10^{-32} s after the Big Bang, and this time is often taken to represent the earliest meaningful 'time after the Big Bang'. The temperatures and pressures during the inflationary epoch were both unimaginably high, and from this point on the Universe continued to expand (non-exponentially) and cool rapidly.

2.2 The nature of the expanding Universe

We should at this point consider the nature of the expansion of the Universe in a little more detail. When thinking about the Big Bang, most people picture matter expanding from a point in all directions within an infinite space, as shown in Figure 2.1(a). While this is a convenient mental picture, it is not really correct, as it is space itself that is expanding, not matter within the space. A two-dimensional analogy is to imagine living on the (two-dimensional) surface of an inflating balloon. In this scenario, the only directions you know are left, right, forwards, and backwards; the concepts

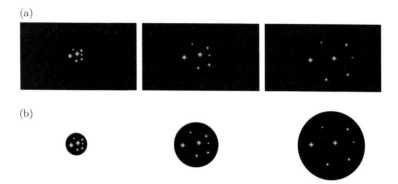

Fig. 2.1 In the expansion of the Universe, matter does not expand into an existing space, as shown in (a), but instead space itself expands at all points simultaneously, as shown in (b).

of 'up' and 'down' are meaningless on a two-dimensional surface. Over time, you observe that objects at rest with respect to the surface of the balloon are in fact moving apart from each other. This is despite the fact that you have explored your entire two-dimensional world and not found any edge or 'outside' for it to expand into. In fact, the entire 'balloon Universe' is expanding as the balloon is inflated, and the observation that objects are moving apart is a result of the space between them (i.e. the surface of the balloon) expanding.

In our three-dimensional Universe, we have known for some time that other galaxies appear to be receding from our galaxy. However, the galaxies are not really travelling through the space away from us, like fragments from a 'Big Bang bomb'; instead, the space between the galaxies and us is expanding, in the same way as in the two-dimensional 'balloon Universe', and this expansion is happening simultaneously at every position in the Universe.[1] This is shown schematically in Figure 2.1(b). Note that since the total energy in the Universe is constant, as the volume of the Universe increases, the energy density, and therefore the temperature, must decrease as the Universe expands.

For many years it was thought that the expansion of the Universe would gradually slow as a result of gravitational attraction between its constituents, perhaps even eventually reversing and leading to a 'Big Crunch'.

---

[1] This analogy was taken from the excellent article *Misconceptions about the Big Bang*, from the March 2005 edition of *Scientific American*.

While repeated cycles of Big Bangs and Big Crunches offer an appealing solution to the question of 'What happened before the Big Bang', the model was disproved in the late 1990s by two independent comprehensive surveys of Type IA supernovae, both originally carried out in order to chart the expansion of the Universe.[2] The surveys showed that the expansion had indeed been slowing, as predicted in the simple model above, until around 5 billion years ago. However, since then, the expansion has been accelerating. Saul Perlmutter, Brian Schmidt, and Adam Riess received the Nobel Prize in Physics for this discovery in 2011. The accelerating expansion of the Universe is thought to be a consequence of the presence of *dark energy*. Dark energy is an intrinsic fundamental energy possessed by a volume of space, also sometimes known as *vacuum energy*, i.e. the energy density of empty space. It is usually quantified in terms of the *cosmological constant*, Λ, estimated to be around 10^{-29} g cm^{-3} (note that energy and mass are equivalent through Einstein's mass–energy relationship $E = mc^2$, discussed further in Section 2.3, hence the units of mass per unit volume). This is a very low density, but because dark energy is evenly distributed through space, it accounts for the majority (68.3%) of the mass–energy balance within the Universe.[3] The remainder is thought to consist of 4.9% ordinary matter, and 26.8% dark matter — an as-yet uncharacterised form of matter that does not absorb or emit electromagnetic radiation, but has clearly observable gravitational effects on ordinary matter, radiation, and the structure of the Universe.

2.3 The first particles

Immediately after the inflationary epoch, the Universe had cooled sufficiently for the first particles to form. These particles formed out of pure energy. Einstein first demonstrated the equivalence and interchangeability of matter and energy in deriving his famous equation, $E = mc^2$. To put this equation into context, creation of 1 gram of mass requires an energy

---

[2]A. G. Riess *et al.* (Supernova Search Team), Observational evidence from supernovae for an accelerating Universe and a cosmological constant, *Astronomical J.*, **116**(3), 1009–1038 (1998); S. Perlmutter *et al.* (Supernova Cosmology Project), Measurements of Omega and Lambda from 42 high red-shift supernovae, *Astrophys. J.*, **517**(2), 565–586 (1999).

[3]P. A. R. Ade, N. Aghanim, C. Armitage-Caplan *et al.* (Planck Collaboration), Planck 2013 Results. I. Overview of products and scientific results, *Astron. Astrophys.*, **1303**, 5062 (2013).

of 89 TeraJoules (8.9×10^{13} J), an almost incomprehensibly large amount of energy. However, fundamental particles weigh only a tiny fraction of a gram, and their equivalent energies, known as their *rest energies*, are measured in MeV (mega electron volts, with 1 MeV = 1.602×10^{-13} J). The rest energies of electrons, protons, and neutrons, the basic constituents of atoms, are 0.511, 938, and 950 MeV, respectively.

At 10^{-32} s after the Big Bang, the first elementary particles began to form. These included quarks, gluons, electrons, and neutrinos. However, with the temperature still a toasty 10^{27} K, it was impossible to create composite particles such as protons and neutrons. At this stage, the Universe consisted of a unique phase of matter known as a *quark-gluon plasma*. In recent years, physicists have put a great deal of effort into trying to recreate this particularly fascinating state of matter, in which the strong interaction is overcome and free quarks and gluons are observed. As far as we know, this state of matter has only ever existed in nature in the instant after the Big Bang, but it may be possible to recreate it in high-energy collision experiments such as those being carried out in the Large Hadron Collider at CERN (Conseil Européen pour la Recherche Nucléaire, which translates as the European Organisation for Nuclear Research).

Around 1 microsecond (10^{-6} s) after the Big Bang, the temperature had cooled to around 10^{13} K, a temperature low enough that quarks could combine to form protons and neutrons. There are six different varieties of quarks (up, down, top, bottom, charm, and strange), but only up and down quarks are involved in the formation of protons and neutrons. A proton is formed from two up quarks and a down quark, while a neutron contains two down quarks and an up quark. Up and down quarks have charges of $+2/3$ and $-1/3$, respectively, giving protons and neutrons their familiar overall charges of $+1$ and 0. Quarks also have a quantum number known as 'colour charge', which can take the values 'red', 'green', or 'blue'. While charge determines how particles will interact via the electromagnetic interaction, colour charge is required to explain how particles interact via the strong interaction. In particular, to satisfy the Pauli's principle that no two identical particles in a system can have the same set of quantum numbers, protons and neutrons must contain one quark of each colour. This also satisfies the principle that hadrons (composite particles made up of quarks) must be colourless (red + green + blue = colourless).

At this stage, a microsecond after the Big Bang, having formed electrons, protons, and neutrons, the fledgling Universe contains all of the building blocks of matter.

2.4 Hydrogen and helium nuclei

Around 100 s after the Big Bang, at a temperature of around 1 billion K, hydrogen nuclei (protons) already existed, and the Universe had cooled enough for the formation of the first composite atomic nuclei. Protons and neutrons reacted to form deuterium nuclei.

$$n + p \rightarrow {}^2\text{H} + \gamma \tag{2.1}$$

where γ is a high energy photon ('gamma particle'). Once deuterium nuclei had formed, there were various pathways to ^{3}He and ^{4}He, all involving addition of one or more protons and neutrons to the nucleus.

$$^2\text{H} + n \rightarrow {}^3\text{H} + \gamma$$
$$^3\text{H} + p \rightarrow {}^4\text{He} + \gamma$$

or

$$^2\text{H} + p \rightarrow {}^3\text{He} + \gamma$$
$$^3\text{He} + n \rightarrow {}^4\text{He} + \gamma$$

2.5 The first atoms

The newborn Universe evolved from a state in which it comprised pure energy to one in which the first atomic nuclei had formed on a timescale of less than two minutes. However, following this period of rapid progress, the Universe went through rather a quiet spell, with nothing remarkable happening for hundreds of thousands of years. During this time the Universe consisted largely of a white hot opaque fog of hydrogen plasma, which was slowly cooling as it expanded. After around 370,000 years, the temperature had reduced to a balmy 10,000 K, sufficiently 'cool' that hydrogen and helium nuclei could capture electrons to form the first neutral atoms. These atoms could no longer absorb the surrounding thermal radiation, and the Universe became transparent. The photons that existed in the Universe at that time have gradually lost energy and become more dispersed as the Universe has continued to expand and cool, but still exist today everywhere in the Universe as the cosmic microwave background radiation. The radiation takes the form of a black-body spectrum with a temperature of 2.725 K, and was first observed in 1964 by Arno Penzias and Robert Wilson,[4] two American radio astronomers who received the Nobel prize for their work in 1978.

Up to this point, the Universe has developed relatively rapidly, forming the first chemical elements spontaneously simply as a consequence of expanding and cooling. However, there is now a real problem to be overcome, in that there are no further stable nuclei that can be formed by neutron capture. Synthesis of heavier elements requires formation of the first stars.

2.6 Questions

2.6.1 Essay-style questions

Q2.1 Explain why the expansion and cooling of the Universe led to the formation of fundamental particles (electrons, muons, tauons, quarks, etc.), followed some time later by composite particles (protons, neutrons, other hadrons).

Q2.2 Explain why, even though the first atomic nuclei formed within around 100 s of the Big Bang, the first neutral atoms did not form for more than 300,000 years.

2.6.2 Problems

P2.1 *Particle rest energies and masses*
The rest energies of an electron, proton, and neutron are 0.511, 938, and 950 MeV, respectively. Determine their masses.

P2.2 *Quarks and subatomic particles*
Protons and neutrons are each composed of three fundamental particles known as quarks. There are six types (or 'flavours') of quark (up, down, top, bottom, charm, and strange), but only up and down quarks are present in protons and neutrons. Up quarks have a charge of $+2/3$, and down quarks have a charge of $-1/3$. In addition, quarks have a 'colour' quantum number, which may be 'red', 'green', or 'blue'. All

---

[4] A. A. Penzias and R. W. Wilson, A measurement of excess antenna temperature at 4080 Mc/s, *Astrophy. J. Lett.*, **142**, 419–421 (1965); A. A. penzias and R. W. Wilson, A measurement of the flux density of CAS A at 4080 Mc/s, *Astrophys. J. Lett.*, **142**, 1149–1151 (1965).

composite particles must be 'colourless', and must therefore be made up either of a quark of each colour, or a quark and its antiquark.

(a) Use the information above to determine the quark combinations within protons and neutrons.
(b) Based on the information given above, what other combinations of two and three quarks can form composite particles? What are their overall charges?

(*Note*: The above is a simplified description of particle formation. The interested reader may like to investigate the properties of quarks and the rules governing particle formation in more detail).

P2.3 *Ionisation temperatures*

The ratio of ionised to neutral H atoms at temperature T is given by the Saha equation

$$\frac{n_{\text{ion}}}{n_{\text{neutral}}} = \frac{1}{n_e} \left(\frac{2\pi m_e k_B T}{h^2} \right)^{3/2} \exp(-I/k_B T)$$

where n_e is the electron number density, m_e is the electron mass, and I is the ionisation potential of atomic hydrogen, 13.5984 eV.

(a) Assuming a total number density of H and H^+ of N, and taking advantage of charge conservation, show that this can be rewritten

$$\frac{x^2}{1-x} = \frac{1}{N} \left(\frac{2\pi m_e k_B T}{h^2} \right)^{3/2} \exp(-I/k_B T)$$

where $x = n_{\text{ion}}/N$ is the ionisation fraction.

(b) The above is a quadratic equation in x. Solve to find the ionisation fraction x, and plot x as a function of T to determine the temperature at which 50% of the hydrogen gas is ionised. Assume $N = 10^{20}$ m^{-3}, typical of a stellar atmosphere.

Chapter 3

Stars and the Creation of the Higher Elements

3.1 Star formation and the nucleosynthesis of heavier elements

In the previous chapter, we looked at the gas-phase processes involved in formation of the first two chemical elements, hydrogen and helium. More hydrogen and helium formed as the Universe continued to cool. By the time the temperature had reduced to around 500 K, somewhere between 200 million and 1 billion years after the Big Bang, hydrogen and helium began to coalesce under the force of gravity into giant gas clouds. Once the gas clouds became large enough (thousands to tens of thousands of solar masses), they started to collapse under the influence of gravity. It is thought that molecular hydrogen also formed at this stage via gas-phase reactions, and that the presence of H_2 was important in driving the gravitational collapse.[1] During gravitational collapse of a gas cloud to form a star, gravitational potential energy is converted to kinetic energy. As the velocity of the particles increases, so does their temperature. In the early stages of star formation, energy can be lost from the gas particles by emission of infrared radiation. However, at some point the gas density increases to the point where the gas becomes opaque to radiation and the temperature increases, firstly to the point at which hydrogen atoms are ionised, and finally to temperatures at which the collisions between the resulting protons have sufficient energy

---

[1] In the modern Universe, H_2 formation is dominated by chemistry occurring on the surface of interstellar dust grains, with gas-phase processes making only a very small contribution.

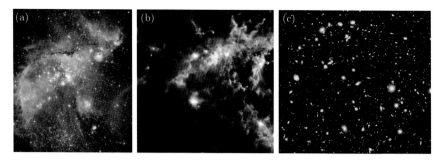

Fig. 3.1 (a) Star-forming region in the Small Magellanic Cloud, 210,000 light years from Earth (NASA image SPD-HUBBLE-STScI-2005-35a); (b) Star-forming region associated with the Rosette Nebula, a stellar nursery in the Monoceros constellation, 5000 light years from Earth (NASA image 450093main_hobys_rosette_05-A3_3d); (c) Hubble ultra-deep field image showing a spectacular array of galaxies within an area of sky 3.4 arc minutes in diameter (created for NASA by the Space Telescope Science Institute under Contract NAS5-26555).

to induce nuclear fusion, and the star begins to 'shine'. Two examples of star-forming regions or 'stellar nurseries' within large gas clouds are shown in Figures 3.1(a) and 3.1(b). As these images indicate, many stars can form within a single giant gas cloud. The stars are held together by their mutual gravity to form a galaxy. The total angular momentum associated with rotational motion of the original gas cloud must be conserved (see Section 5.4.9), with the result that galaxies tend to form elliptical or spiral shapes, each star orbiting around the centre of mass of the cloud. Tens of thousands of galaxies have now been catalogued, each containing between 10^7 and 10^{14} stars. Figure 3.1(c) shows an ultra-deep-field image taken with the Hubble space telescope of a tiny patch of sky corresponding to an angular diameter around one-tenth that of a full Moon as viewed from Earth. The image covers the full detectable range of ultraviolet (UV) to infrared light, and reveals thousands upon thousands of galaxies of stars. The light imaged from some of these galaxies was emitted only around 400–800 million years after the Big Bang, around 13.2 billion years ago.

The first stars, known as Population III stars, were huge, hundreds of times heavier than our Sun, and burned out relatively quickly compared to later stars, on a timescale of 3–4 million years. Perhaps somewhat counterintuitively, more massive stars have shorter lifetimes. Although the amount of fuel a star possesses increases with mass, m, the rate at which it consumes this fuel increases even faster. When all factors are taken into account, it is found that a star's lifetime on the main sequence (the period when it is burning hydrogen as fuel, described in more detail in the following) is proportional to $1/m^3$.

For most of their lifetime, the Population III stars were fuelled largely by *hydrogen burning*, a multistep process in which four ^{1}H nuclei are fused to form a single ^{4}He nucleus and various other elementary particles. There are two different mechanisms for hydrogen burning, but in relatively small stars, including our Sun, the *proton–proton cycle* dominates, and produces around 90% of the star's energy.[2]

$$\begin{aligned}^1\mathrm{H} + {}^1\mathrm{H} &\rightarrow {}^2\mathrm{H} + e^+ + \nu_e \\ {}^2\mathrm{H} + {}^1\mathrm{H} &\rightarrow {}^3\mathrm{He} + \gamma \\ {}^3\mathrm{He} + {}^3\mathrm{He} &\rightarrow {}^4\mathrm{He} + 2{}^1\mathrm{H}\end{aligned} \qquad (3.1)$$

Once the supply of ^{1}H in the core of a star is exhausted, near the end of its lifetime, the dominant process becomes *helium burning*, the fusion of two ^{4}He nuclei to form a ^{8}Be nucleus.

$$^4\mathrm{He} + {}^4\mathrm{He} \rightarrow {}^8\mathrm{Be} + \gamma \qquad (3.2)$$

The ^{8}Be nucleus resulting from this process is not stable, and usually decays back to two ^{4}He nuclei, but can also undergo other reactions. For example,

$$^8\mathrm{Be} + {}^4\mathrm{He} \rightarrow {}^{12}\mathrm{C} + \gamma \qquad (3.3)$$

From ^{12}C, it is possible to make all of the even numbered elements up to ^{56}Fe through analogous nuclear fusion processes:

$$\begin{aligned}{}^{12}\mathrm{C} + {}^4\mathrm{He} &\rightarrow {}^{16}\mathrm{O} + \gamma \\ {}^{16}\mathrm{O} + {}^4\mathrm{He} &\rightarrow {}^{20}\mathrm{Ne} + \gamma \\ {}^{20}\mathrm{Ne} + {}^4\mathrm{He} &\rightarrow {}^{24}\mathrm{Mg} + \gamma \\ {}^{24}\mathrm{Mg} + {}^4\mathrm{He} &\rightarrow {}^{28}\mathrm{Si} + \gamma\end{aligned} \qquad (3.4)$$

and so on. Odd numbered nuclei are formed in less efficient reactions involving proton capture. Many of the original Population III stars probably did not proceed much further than helium burning before reaching the end of their lifetime, at which point they exploded in supernovae, scattering the elements they had formed through space. As far as we know, no Population III stars are still in existence.

---

[2]In stars with masses more than about 1.5 times the mass of the Sun, the carbon–nitrogen–oxygen (CNO) cycle dominates, in which carbon, nitrogen, and oxygen nuclei are involved in the sequence of steps that converts hydrogen to helium.

Over time, the cycle of giant gas cloud formation and star formation was repeated to form a new generation of stars known as Population II stars. The presence of heavier elements in the gas clouds modified their cooling and contraction properties, with the result that Population II stars tend to be smaller than the first generation of stars, with much longer lifetimes. Many Population II stars are still in existence, and are thought to be between 10 and 13 billion years old (for comparison, remember that the Universe is thought to be around 13.8 billion years old). These stars have a much higher fraction of 'heavy' (i.e. non-hydrogen and helium) nuclei than the Population III stars.

Many Population II stars have reached the end of their lives, and the star-formation cycle has repeated itself to form Population I stars. These have ages of less than 10 billion years, and contain even higher levels of heavy nuclei, having been given a 'head start' by their 'parent' Population II stars. Our Sun is a Population I star. To keep things in perspective, even in a so-called 'high metallicity' star such as the Sun, only around 1.8% of nuclei within the Sun are heavy elements (the *metallicity* of a star is the proportion of its matter made up from elements other than H and He).

Nuclear fusion processes of the type described above are energetically favourable until the ^{56}Fe nucleus is reached. Iron has the most stable nucleus in terms of binding energy per nucleon, as shown in Figure 3.2. Beyond ^{56}Fe, fusion processes are no longer energetically favourable, and heavier elements are instead built up gradually through a combination of neutron capture

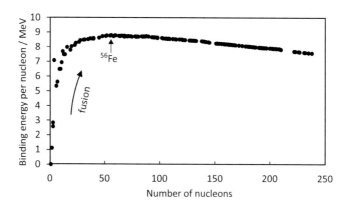

Fig. 3.2 Binding energy per nucleon for the chemical elements.
Source: Data from NUBASE (see G. Audi, O. Bersillon, J. Blachot, and A. H. Wapstra, The NUBASE evaluation of nuclear and decay properties, *Nucl. Phys. A*, **729**, 3–128 (2003)).

and β decay (electron emission). In the example below, ^{98}Mo (atomic number 42) captures a neutron to form ^{99}Mo, which then emits an electron to form ^{99}Tc, with atomic number 43, and a neutrino. Neutron capture followed by electron emission therefore has the overall effect of increasing the atomic number by one.

$$^{98}\text{Mo} + n \rightarrow {}^{99}\text{Mo} + \gamma$$
$$^{99}\text{Mo} \rightarrow {}^{99}\text{Tc} + \text{e}^- + \nu \tag{3.5}$$

3.2 Dispersion of the chemical elements into interstellar space

The dispersion of chemical elements from stars into the interstellar medium occurs at the end of a star's lifetime, with the dispersion mechanism depending on the mass of the star. The life cycles of a typical low-mass (less than three solar masses) and high-mass star are shown schematically in Figure 3.3.

During the time when a star is fuelled primarily by hydrogen burning it is known as a *main sequence star*. For a star similar in size to our Sun, this period lasts around 10 billion years. Once all the hydrogen in the core has

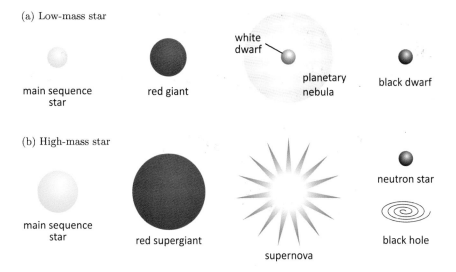

Fig. 3.3   Life cycles of typical low-mass and high-mass stars.

been converted to helium, the hydrogen fusion reactions in the core stop. However, they continue in the shell around the core. The core begins to cool and contract under gravity. Eventually, in a sequence of processes similar to the formation of the original star, the core of the star has condensed and increased in temperature to the point where the density and temperature are high enough to initiate helium burning. The outer layers of the star begin to cool, expand, and shine less brightly, and the star is now a *red giant*. Eventually, the helium within the star's core is converted to carbon, at which point the core becomes a *white dwarf*, and the outer layers of the star drift away to form a gaseous shell called a *planetary nebula*. The atomic species within the planetary nebula will eventually drift away into the interstellar medium. Once nuclear fusion reactions stop completely, the core of the now dead star is known as a *black dwarf*. Some low-mass stars at various points in their life cycles are shown in Figure 3.4.

The sequence of events unfolds rather differently for massive stars, 3–50 times heavier than our Sun. Their main-sequence lifetimes are much shorter, at millions rather than billions of years. Once a massive star has exhausted its supply of hydrogen and begins helium burning, it expands much more than a low-mass star, forming a *red supergiant*. An example of a red supergiant that is clearly visible in the night sky is Betelgeuse,

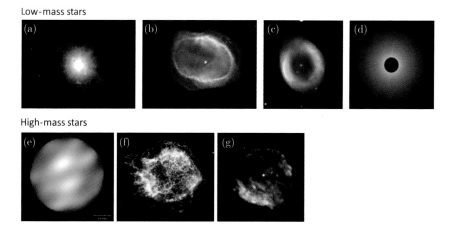

Fig. 3.4 NASA images of low-mass and high-mass stars at various points in their life cycles: (a) Mira, a red giant; (b) NGC6751 white dwarf and planetary nebula; (c) M57 ring nebula (and white dwarf); (d) the shadow cast by a black dwarf; (e) Betelgeuse, a red supergiant; (f) and (g) Cassiopeia A, a remnant of a supernova 11,000 years ago — the neutron star is visible in the centre.

shown in Figure 3.4. Over the next million years after becoming a red supergiant, many different nuclear reactions occur in the core of the star, forming different elements in shells around an iron core. Eventually, the star runs out of fuel, leading to a spectacular gravitational collapse and subsequent explosion called a supernova. In contrast to the millions of years over which the star has evolved up to this point, the supernova event often occurs on a timescale of seconds. The resulting shock wave blows off the outer layers of the star into the interstellar medium, but sometimes the core will survive the explosion. If the remaining core has a mass of between 1.5 and 3 solar masses, it will contract to form a very dense *neutron star*, while a higher-mass core will contract to form a *black hole*.

3.3 Cosmic abundance of the elements

All of the chemical elements in the Universe today formed through nuclear processes within stars of the type described above. The cosmic abundances of the different elements are the result of a complex interplay between the various formation mechanisms operating in stellar cores, and show considerable variability from element to element. The cosmic abundances of all naturally occurring chemical elements, from hydrogen to uranium, are plotted in Figure 3.5 (note the log scale). The plot shows several notable features:

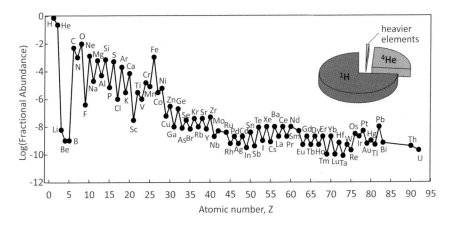

Fig. 3.5 The cosmic abundance of the elements.

(1) The Universe consists of around 74% hydrogen, 24% helium, and only 2% heavier elements, a result of the facile formation of H and He (described in Chapter 2) even in the absence of star formation.
(2) The abundance falls off from light elements to heavier elements, reflecting the fact that in all formation mechanisms, heavier elements are built up gradually from nuclear reactions of lighter elements.
(3) Iron has the most stable nucleus, and consequently is one of the most abundant elements.
(4) Lithium, beryllium, and boron are extremely rare compared with other elements, since there is no straightforward way to synthesise them through nuclear reactions.
(5) Even elements are more abundant than odd since their nuclei are more stable and they are formed more efficiently.

Once appreciable quantities of common nuclei such as carbon, nitrogen, and oxygen had built up in the interstellar medium, the first molecules began to form via a variety of chemical mechanisms. The chemical conditions in the interstellar medium, and the rich chemistry occurring in this rather unfriendly chemical environment, are the subject of Chapter 4.

3.4 Questions

3.4.1 Essay-style questions

Q3.1 Discuss the events leading to star formation within a molecular cloud.
Q3.2 Explain the main features of the cosmic abundance plot in Figure 3.5.

3.4.2 Problems

P3.1 *Size of an interstellar gas cloud*
The mass of the Sun is 2×10^{30} kg. Calculate the smallest diameter of a (spherical) interstellar gas cloud with a density of $1000 \, \text{cm}^{-3}$ that could collapse to form a star with one solar mass. Give your answer in units of (a) metres and (b) light years.

P3.2 *A very simple model of proton–proton fusion*
In order for nuclear fusion to occur, two protons must reach separations at which the attractive strong nuclear force overcomes

the Coulomb repulsion between the two positive charges.

(a) Assuming that this occurs at a range of around 1 fm (1×10^{-15} m), calculate the approximate barrier to nuclear fusion, assuming that the only contribution is from Coulomb repulsion.

(b) Assuming a head-on collision between two protons with equal and opposite momenta, what is the minimum velocity at which each proton must be travelling in order to achieve fusion?

Hint: Start by determining the minimum kinetic energy each proton must possess.

P3.3 The intensity $I(\lambda, T)$ of radiation emitted from a star with surface temperature T as a function of wavelength λ can be modelled by Planck's law of black-body radiation.

$$I(\lambda, T) = \frac{2hc^2}{\lambda^5} \frac{1}{\exp(hc/\lambda k_\mathrm{B} T) - 1}$$

On the same set of axes, plot the intensity distribution of the radiation emitted from the Sun (surface temperature 5778 K) and Betelgeuse (surface temperature 3250 K) as a function of wavelength. Use your plots to predict the observed colours and brightnesses of the two stars.

Chapter 4

Interstellar Chemistry — Molecules in Space

4.1 The interstellar medium

Most of the chemistry occurring in space occurs within the interstellar medium, the regions between stars within a galaxy. In the vast intergalactic spaces between galaxies, with only around one hydrogen atom per cubic metre, the density is simply too low for any chemistry to occur. In contrast, within the interstellar medium, densities are between 1 and 10^6 particles (atoms or molecules) per cubic centimetre,[1] sufficient for a rich variety of chemical reactions to occur, albeit over a long timescale (for comparison, the molecular density at standard temperature and pressure is around $3 \times 10^{19}\,\text{cm}^{-3}$), leading to the formation of many different molecular species.

The interstellar medium is often subdivided into a number of different environments, depending on the number density, temperature, and atomic and molecular composition. These are detailed below.

4.1.1 Diffuse interstellar medium

The diffuse interstellar medium is effectively 'empty space'. Particle densities are around $1\text{--}100\,\text{cm}^{-3}$, and while the translational temperatures of

---

[1] Note that number densities, i.e. the number of particles per unit volume, have dimensions of a pure number (dimensionless) divided by a volume, i.e. overall dimensions of inverse volume. Number densities are therefore typically given in units of cm^{-3} or m^{-3} rather than the perhaps more intuitive 'particles. cm^{-3}'.

atoms and molecules may be as high as 100 K, the concept of temperature is not really applicable with such low number densities. These regions of the interstellar medium contain mostly atomic hydrogen and atomic ions, with a small molecular fraction (defined by the ratio of H_2 to H) of between 0.0 and 0.1.

4.1.2 Giant molecular clouds

Giant molecular clouds are enormous collections of gas that can be tens of light years in diameter and contain most of the mass within the interstellar medium. As explained in the previous chapter, these are the regions of new star formation. Some images of giant molecular clouds from the NASA image gallery are shown in Figure 4.1. The average density within a giant molecular cloud is around 100–1000 cm^{-3}, but the most dense regions may have densities of up to 10^6 cm^{-3}. Temperatures range from around 100 K near the edges of a cloud to around 10 K in the centre. As the name suggests, giant molecular clouds are home to a considerable amount of chemistry, and contain many different molecular species (see Table 4.1), over 160 of which have been definitively identified. Even so, it is worth noting that the average density within such a cloud is comparable to the best vacuum achievable

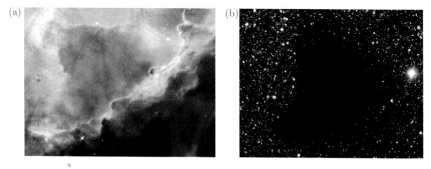

Fig. 4.1 Giant molecular clouds: (a) a region of the M17 molecular cloud within the Omega nebula of Sagittarius. The Sagittarius constellation is around 5500 light years away from Earth, and the image spans around 3 light years. Green, red, and blue hues correspond to abundant hydrogen and trace amounts of sulphur and oxygen, respectively; (b) the Barnard 68 molecular cloud. The cloud appears dark against the background of the stars, as visible light is unable to penetrate through the cloud due to absorption and scattering by hydrogen and dust grains (NASA image gallery).

Table 4.1 Molecular species identified in giant molecular clouds. Unconfirmed detections are denoted with a question mark.

Diatomic molecules	AlF, AlCl, C_2, CH, CN, CO, CP, CS, CSi, HCl, H_2, KCl, NH, NO, NS, NaCl, OH, PN, SO, SiN, SiO, SiS, HF, SH, SiH, CH^+, CD^+, CO^+, SO^+
Triatomic molecules	C_3, C_2H, C_2O, C_2S, CH_2, HCN, HCO, H_2O, H_2S, HNC, HNO, MgCN, MgNC, N_2O, NaCN, OCS, SO_2, c-SiC_2, CO_2, NH_2, SiCN, HCO^+, DCO^+, HCS^+, HOC^+, N_2H^+, H_3^+, H_2D^+
Four-atom molecules	c-C_3H, l-C_3H, C_3N, C_3O, C_3S, C_2H_2, HCCN, HNCO, HNCS, H_2CO, H_2CN, H_2CS, NH_3, SiC_3, CH_3, CH_2D^+?, $HCNH^+$, $HOCO^+$, H_3O^+
Five-atom molecules	C_5, C_4H, C_4Si, l-C_3H_2, c-C_3H_2, CH_2CN, CH_4, HC_3N, HC_2NC, HCOOH, H_2CHN, H_2C_2O, H_2NCN, HNC_3, SiH, H_2COH^+
Six-atom molecules	C_5H, C_5O, C_2H_4, CH_3CN, CH_3NC, CH_3OH, CH_3SH, HC_3NH^+, HC_2CHO, $HCONH_2$, l-H_2C_4, C_5N, HC_3NH^+
Seven-atom molecules	C_6H, CH_2CHCN, CH_3C_2H, HC_5N, $HCOCH_3$, NH_2CH_3, c-C_2H_4O
Eight-atom molecules	CH_3C_3N, $HCOOCH_3$, CH_3COOH?, CH_2OHCHO, H_2C_6, C_7H
Nine-atom molecules	CH_3C_4N, CH_3CH_2CN, $(CH_3)_2O$, CH_3CH_2OH, HC_7N, C_8H
10-atom molecules	CH_3C_5N?, $(CH_3)_2CO$, NH_2CH_2COOH?
11-atom molecules	HC_9N
13-atom molecules	$HC_{11}N$

Source: Data from *Molecules in Space*, by J. Tennyson, Volume 3, Part 3, Chapter 14 of the *Handbook of Molecular Physics and Quantum Chemistry* (John Wiley and Sons, 2003).

on Earth. Molecular clouds are often classified into subcategories as follows:

(1) Diffuse molecular clouds: most matter is present in atomic, rather than molecular form. The molecular fraction (defined as the ratio of the number of H_2 molecules to the total number of H atoms) lies between 0.1 and 0.5;

(2) Translucent molecular clouds: most H is present in molecular form, but most carbon is present in atomic form as C or C^+, with some present in molecular form, predominantly as CO;

(3) Dense molecular clouds: virtually all H is present as H_2, and most of the carbon is also present in molecular form. Dense clouds consist primarily of H_2, with CO the next most abundant molecule, at around 10^{-4} times the number density of H_2. Other simple molecules such as CH, OH, CN,

and H_2O are present at considerably lower number densities, totalling around 10^{-7} of the H_2 density.

4.1.3 Circumstellar medium

As the name suggests, the circumstellar medium encapsulates the regions directly around a star. The environment within these regions depends on the type and extent of evolution of the star. Regions around young stars experience high photon fluxes in the ultraviolet (UV) so that all molecules are photodissociated and photoionised; such regions are sometimes referred to as photon-dominated regions. Around older stars there may be significant dust, leading to surface chemistry as well as scattered starlight.

For the remainder of this chapter, we will focus our attention on giant molecular clouds. The atomic composition of these regions is determined by the past history of nearby stars (see Sections 3.1 and 3.2), which eject processed nuclear material via stellar winds and supernova explosions. As we shall see, the molecular composition reflects the balance between chemical evolution via reactions, destruction of molecules by light from stars or by cosmic rays, and condensation and subsequent reaction on dust grains.

4.2 Chemistry in interstellar space

Molecular clouds are characterised by very low temperatures, on the order of 10 K. At these temperatures there is insufficient energy for collisions to overcome any activation barrier to reaction, and therefore the only gas-phase chemical reactions that can proceed at such low temperatures are radical–radical reactions and ion–molecule reactions, both of which are barrierless[2] and proceed on every collision. Interstellar gas clouds also have exceptionally low densities by terrestrial standards, resulting in extremely low collision frequencies and relatively few opportunities for chemistry to occur. The collision frequency is simply the number of collisions occurring

---

[2] Note that though these reactions are barrierless in the sense that there is no activation barrier on the reaction potential energy surface (PES) along the reaction coordinate, as we shall see in Section 5.4.19, such reactions may have a *centrifugal barrier* arising from conservation of angular momentum during the reactive collision, and this can have a significant effect on the reaction rate.

within a unit volume each second, and is given by

$$z = \sigma_c \bar{v}_{\text{rel}} n_A n_B \qquad (4.1)$$

where σ_c is the collision cross-section (see Section 5.4.6 for more details), $\bar{v}_{\text{rel}}$ is the mean relative velocity of the collision partners, and n_A and n_B are the number densities (number of atoms or molecules per unit volume) of the species of interest. For a system at thermal equilibrium, which is not always the case at the very low number densities found in many regions of interstellar space, each chemical species has a Maxwell–Boltzmann distribution of velocities, and the mean relative velocity is given by

$$\bar{v}_{\text{rel}} = \left(\frac{8 k_B T}{\pi \mu} \right)^{1/2} \qquad (4.2)$$

where T is the temperature, μ is the reduced mass, and k_B is Boltzmann's constant. Even in the densest regions of a molecular cloud, with number densities of $10^6 \, \text{cm}^{-3}$, collision rates are around $5 \times 10^{-4} \, \text{s}^{-1}$, or approximately one collision every half an hour. In less dense regions, atoms and molecules may go for many weeks, or even longer, between collisions. Chemistry therefore occurs at a very slow rate in interstellar space when compared with the timescales we are used to on Earth. However, since giant molecular clouds last for around 10–100 million years before they are dissipated by heat and stellar winds from stars forming within them, there is plenty of time for some quite complex chemistry to occur, albeit at a rather leisurely rate.

The very low collision frequency has important consequences for the types of molecules that may form in interstellar space. As can be seen from Table 4.1, terrestrial concepts of molecular stability simply do not apply in this extremely non-reactive environment. Carbon does not need to have four bonds; in fact, there are many subvalent species, radicals, molecular ions and energetic isomers amongst the molecules observed in interstellar gas clouds. Carbon-containing compounds tend to be highly unsaturated, with many double and triple bonds, and few branched chains. Polyynes are commonly observed, some with fairly long chain lengths, for example H–C≡C–C≡C–C≡C–C≡C–C≡C–C≡N. Many of the molecules observed in space would react almost instantly were they to be transported to Earth.

The extremely infrequent collisions between atoms and molecules render a number of terrestrial chemical principles, for example the idea of 'thermalisation', irrelevant. While thermal rate constants are often quoted for reactions occurring in interstellar space, these are generally the result of laboratory-based measurements made on Earth under conditions for which thermalisation does occur. For the most part, when considering chemistry in the interstellar medium, rather than employing the terminology of 'bulk' chemical kinetics, it generally makes more sense to consider such reactions in terms of individual collisions. We will return to this idea in detail when considering theoretical studies of astrochemical reactions in Chapter 5. Note that an overview of the basic principles of chemical kinetics is provided in Appendix A.

4.3 Molecular synthesis in interstellar gas clouds

Gas-phase molecular synthesis in interstellar clouds is believed to occur primarily via ion–molecule reactions, with some contribution from neutral reactions. Since the molecular species identified from spectroscopic data are mostly neutral, the ionic species formed as a result of ion–molecule chemistry must become charge-neutral relatively quickly. Molecular synthesis can therefore be described by the following general scheme.

$$\text{Neutral gas} \xrightarrow{\text{ionisation}} \text{Small ions} \xrightarrow{\text{reaction}} \text{Large ions} \xrightarrow{\text{neutralisation}} \text{Observed species} \tag{4.3}$$

Chemistry can also occur on the surface of dust grains. Surface-catalysed reactions of this type turn out to be very important in the interstellar medium, and we will look at these in some detail later in Section 4.6.3.

We will now consider the types of reaction contributing to each of the three stages involved in gas-phase interstellar chemistry, i.e. ionisation, reaction, and neutralisation, before investigating dust-grain-catalysed chemistry.

4.4 Ionisation processes in the interstellar medium

Photoionisation of atoms and molecules is known to be a very common process in regions of interstellar space close to stars. On first consideration, photoionisation may also appear to be a prime candidate to provide the dominant ionisation mechanism within molecular clouds. However, the high density of hydrogen and dust grains in molecular clouds prevents visible and UV light from penetrating very far into the interior. Clear evidence for this is provided by the fact that molecular clouds often appear dark when viewed through a telescope, since they block the light from stars behind them. An example is the Barnard 68 molecular cloud shown in Figure 4.1(b). In contrast, infrared light can penetrate into molecular clouds (consequently, infrared spectroscopy is a key method for identifying molecular species within these regions), but infrared photons do not have sufficient energy to ionise neutral molecules. Photoionisation can therefore be discounted as an ionisation mechanism to initiate ion–molecule chemistry. Instead, most ions within molecular clouds are formed through collisions with cosmic rays. Cosmic rays are extremely high-energy particles emitted by stars, comprising around 84% protons, 14% alpha particles, and 2% electrons, heavier nuclei, and more exotic particles. Numerous chemical processes can result from a collision of a molecule with a cosmic ray (cr), as summarised below.

$$
\begin{aligned}
AB + cr &\rightarrow AB^+ + e^- + cr &&\text{ionisation} \\
AB + cr &\rightarrow A + B + cr &&\text{dissociation} \\
AB + cr &\rightarrow A + B^+ + e^- + cr &&\text{dissociative ionisation} \\
AB + cr &\rightarrow AB^* + cr &&\text{excitation}
\end{aligned}
$$

Energy is of course conserved in all of these collisions, and consequently, the cosmic rays appearing on the right-hand side of the above equations are lower in energy than those on the left, having given up some of their energy to drive the chemical process of interest.

Formation of negative ions is also possible in interstellar space. Electrons fairly commonly attach to large carbon-based molecules, such as polycyclic aromatic hydrocarbons or PAHs (these large molecules are covered in more detail in Section 4.8.3.3), yielding a negative ion. Sometimes electron attachment is dissociative, in which case the rate can be extremely fast, with rate constants up to $10^{-7}\,\text{cm}^3\,\text{s}^{-1}$. In non-dissociative attachment, emission of a photon will generally be required in order to stabilise the ion, e.g.

$$e^- + \text{PAH} \rightarrow \text{PAH}^- + h\nu \qquad (4.4)$$

4.5 Gas-phase chemical reactions in the interstellar medium

A wide variety of reaction mechanisms operate within the interstellar medium. These can be categorised into bond formation, bond breaking, and rearrangement reactions, as follows:

Bond formation:	Radiative association (neutral or ionic)	$A + B \rightarrow AB + h\nu$
	Associative detachment	$A^- + B \rightarrow AB + e^-$
	Dust-grain-catalysed reaction (neutral or ionic)	$A + B + \text{grain} \rightarrow AB + \text{grain}$
Bond breaking:	Photodissociation	$AB + h\nu \rightarrow A + B$
	Collisional dissociation (neutral or ionic)	$AB + M \rightarrow A + B + M$
	Dissociative recombination	$AB^+ + e^- \rightarrow A + B$
Rearrangement:	Ion–molecule reaction	$A^+ + BC \rightarrow AB^+ + C$
	Charge transfer	$A^+ + B \rightarrow A + B^+$
	Neutral reaction	$A + BC \rightarrow AB + C$

Apart from photodissociation, these processes are all bimolecular, and usually diffusion controlled, with rate constants of around 10^{-9} cm^3 s^{-1}. The various types of reaction are considered in more detail and illustrated with examples in the following.

4.6 Bond-forming reactions

4.6.1 Radiative association

In an association reaction, two reactants combine to form a single product. As a consequence of conservation of energy and linear momentum, the product is usually highly internally excited, and therefore prone to dissociation. In terrestrial chemistry, such reactions normally rely on a subsequent collision with a 'third body' in order to carry away some of the energy of the excited product (while conserving momentum) to yield a stable molecule. The low collision frequency in an interstellar gas cloud rules out this pathway to stabilisation, and instead the highly energised product either decays back to reactants or is stabilised by emission of a photon. For example,

$$C^+ + H_2 \rightleftharpoons CH_2^{+*} \rightarrow CH_2^+ + h\nu$$

$$CH_3 + H_2O \rightarrow CH_3^+.H_2O + h\nu \rightarrow CH_3OH_2^+$$

If the product has an allowed electronic transition back to the ground state, then radiative association can be very efficient; otherwise, it is slow, relying on infrared emission to relax the excited molecule via vibrational relaxation. Rate constants for radiative association vary from $10^{-17}\,\text{cm}^3\,\text{s}^{-1}$ for some diatomics up to $10^{-9}\,\text{cm}^3\,\text{s}^{-1}$ for polyatomics.

Radiative association reactions can build large ions rapidly in a single step. They are difficult to study in the laboratory, and are often probed by studying the collisionally-stabilised analogue. Some examples will be given in Chapter 6.

4.6.2 Associative detachment

In associative detachment, a negative ion and a neutral combine and the resulting negative ion detaches an electron. For example,

$$\text{OH}^- + \text{H} \rightarrow \text{H}_2\text{O} + \text{e}^-$$

This type of reaction is fairly common for small species, and is also thought to occur with larger molecules.

4.6.3 Dust-grain-catalysed reactions

Around 1% of the mass of the interstellar medium consists of dust grains formed in the outflows of dying stars, particularly red supergiants (see Section 3.2). Molecules such as SiO and TiO form in the outer layers of these stars, and the large stellar winds that develop once the star ends its hydrogen-burning phase blow these molecules out into the interstellar medium, where they form aggregation nuclei for dust particles. The dust aggregates into crystalline structures, forming a silicate core a few hundred nanometres in diameter through collection of oxygen atoms from the interstellar medium. Condensation of other molecules from within interstellar gas clouds leads to a layered mantle of ice on the surface, with inner layers containing organic molecules and outer layers containing molecules such as H_2O, CO, CO_2, methanol, H_2CO, NH_3, and so on. A spectrum of interstellar dust (from the W33A dust-embedded massive young star), with some of the identifiable absorption features labelled, is shown in Figure 4.2.

Dust-grain chemistry is difficult to study, and is currently not well understood. However, the mechanisms by which dust-grains-catalyse reactions in space must be similar to those involved in surface catalysis on Earth. Dissociative adsorption to a surface yields highly reactive species

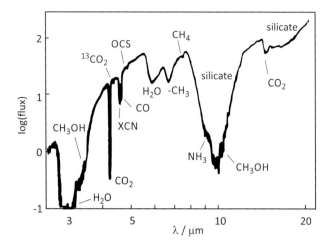

Fig. 4.2 Spectrum of interstellar dust from the W33A dust-embedded massive young star.

Source: Adapted from Gibb *et al.*, *Astrophys. J.*, **536**, 347 (2000).

and alternative reaction pathways, allowing reactions to proceed much more quickly than they would in the gas phase. A reaction of vital importance in the interstellar medium which is known to occur almost exclusively on the surface of dust grains is the formation of H_2 from two H atoms adsorbed to the surface.

$$H + H \xrightarrow{\text{dust grain}} H_2 \qquad (4.5)$$

Organic synthesis is also thought to occur on the surface of dust grains, with adsorption of CO to the surface providing a carbon source to initiate such reactions. For example,

$$CO + H \rightarrow HCO$$
$$HCO + H \rightarrow H_2CO$$
$$H_2CO + H \rightarrow H_3CO$$
$$H_3CO + H \rightarrow CH_3OH \qquad (4.6)$$

4.7 Bond breaking reactions

4.7.1 Photodissociation and collisional dissociation

Both photodissociation and collisional dissociation are common processes on Earth, and occur equally readily in interstellar space. In photodissociation, the energy required to break a chemical bond is provided by a photon, whereas in collisional dissociation the energy is provided by a collision.

4.7.2 Dissociative recombination

In dissociative recombination an ion combines with an electron to produce a highly energised neutral. Since a 'third body' collision to carry away the energy of the neutral is highly unlikely, the product fragments into smaller neutral species. For example,

$$\begin{aligned} H_3O^+ + e^- \rightarrow\ & OH + 2H + 1.3\text{ eV} && (29\%) \\ & OH + H_2 + 5.7\text{ eV} && (36\%) \\ & H + H_2O + 6.14\text{ eV} && (5\%) \\ & O + H + H_2 + 1.4\text{ eV} && (30\%) \end{aligned}$$

As a result of the electrostatic attraction between the ion and the electron and the resulting large collision cross-section, such processes can be extremely fast. Rate constants of up to 10^{-6} cm^3 s^{-1} are common, significantly larger than those for ion–molecule reactions.

4.8 Rearrangement reactions

4.8.1 Charge transfer

Charge transfer involves the transfer of an electron from a neutral to an ion, and may lead to dissociation of the resulting ion. The charge transfer often occurs at large separations of up to 10 Å, and the process has a correspondingly large reaction cross-section.

An example is the dissociative charge transfer from He$^+$ to CO (electron transfer from CO to He$^+$).

$$He^+ + CO \rightarrow C^+ + O + He + 2.2\text{ eV} \tag{4.7}$$

The large ionisation energy of He (24.6 eV) is released during the charge transfer, leading to fragmentation of the product CO$^+$ back into its atomic constituents. This might seem like a backward step in terms of molecular synthesis, but the C$^+$ ion formed in this reaction can and does go on to react further, and is important in building up higher hydrocarbons in interstellar space.

4.8.2 Neutral reactions

Fewer neutral reactions are involved in the interstellar medium than ion-molecule reactions, as they tend to have activation barriers, but even reactions with barriers can be important in high-temperature shocked regions; for example, when a supernova shock wave passes through a gas cloud, the gas is compressed and can heat up to over 1000 K.

4.8.3 Ion–molecule reactions

Ion–molecule reactions are usually barrierless (though there can often be a centrifugal barrier, as we shall see in Section 5.4.19), and make up the majority of bimolecular reactions occurring in the insterstellar medium. The charge-transfer reactions described in Section 4.8.1 are a subset of ion-molecule reactions, but there are many other varieties. A few examples are given below.

4.8.3.1 *Hydrogen atom abstraction*

Hydrogen atom abstraction reactions are important due to the high abundance of atomic hydrogen in the interstellar medium. One of the most common reactions is

$$H_2 + H_2^+ \rightarrow H_3^+ + H + 1.7 \text{ eV} \quad \text{(fast)} \tag{4.8}$$

This is a fast reaction due to the high abundance of both reactants. The H_3^+ ion is responsible for most of the proton-transfer chemistry occurring in the interstellar medium, and is therefore a key molecular reagent in interstellar space.

Another example of a hydrogen atom abstraction is the reaction between NH_3^+ and H_2.

$$NH_3^+ + H_2 \rightarrow NH_4^+ + H + 0.9 \text{ eV} \quad \text{(slow)} \tag{4.9}$$

This reaction occurs much more slowly than the previous reaction, and has an interesting temperature dependence. The reaction involves a barrier, such that the rate slows as the temperature is reduced from room temperature down to around 70 K. At lower temperatures the rate increases again as a quantum tunnelling mechanism takes over. Tunnelling through the activation barrier becomes easier (and therefore faster) at lower temperatures since the collisions are slower and the reactants spend more time in close proximity. Within molecular clouds, the reaction proceeds almost entirely via the tunnelling mechanism.

4.8.3.2 *Proton transfer*

Transfer of a proton from a species of lower to higher proton affinity (PA) is a common process in interstellar space. The reaction exoergicity can be determined from the difference in proton affinities for the species involved. The H_3^+ ion is commonly involved in such reactions. For example,

$$H_3^+ + H_2O \rightarrow H_3O^+ + H_2 + 2.8 \text{ eV} \qquad (4.10)$$

In this case the exoergicity is $\Delta E = \text{PA}(H_2O) - \text{PA}(H_2) = 2.8$ eV.

4.8.3.3 *Carbon insertion*

Carbon insertion reactions involve the insertion of a C^+ ion into a carbon chain. An atom or electron must be ejected during the process in order to conserve momentum, e.g.

$$C^+ + C_2H_2 \rightarrow C_3H^+ + H + 2.2 \text{ eV} \qquad (4.11)$$

Such reactions are important in the synthesis of many of the carbon-containing molecules in the interstellar medium. As an example, the C_3H^+ ion formed in the above reaction can go on to react further, eventually forming cyclic C_3H_2 via a sequence of steps involving hydrogen abstraction and electron–ion recombination.

$$C_3H^+ + H_2 \rightarrow C_3H_2^+ + H$$
$$C_3H_2^+ + e^- \rightarrow C_3H_2 \qquad (4.12)$$

Sequential carbon insertion processes can generate very large organic species known as PAHs. The mechanism for PAH formation in interstellar space is thought to involve a number of steps. First, carbon insertion reactions lead to the formation of carbon chains in a complex series of steps. Further radical–radical and dust-grain-catalysed chemistry then leads to ring formation and continued chain propagation. Some examples of PAH structures are shown in Figure 4.3.

The density of PAHs in the interstellar medium has not been quantified definitively, but may be comparable to that of other simple species such as OH, CH, CN, and H_2O, and they may account for up to 20% of the carbon in these regions.

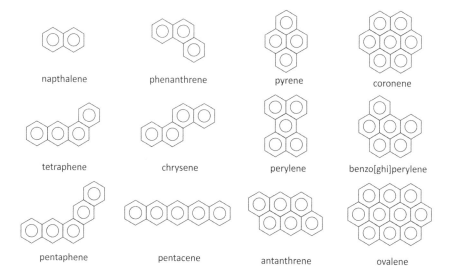

Fig. 4.3 Examples of polycyclic aromatic hydrocarbons.

4.8.3.4 *Rearrangement reactions*

Rearrangement reactions usually occur via a collision complex. A large amount of rearrangement can occur within a highly energised complex, leading to fragmentation to various different sets of products. The fragmentation products often go on to react further, leading to a rich ion chemistry. Consequently, rearrangement reactions are important for coupling different reaction sequences together. A simple example of a rearrangement reaction is the reaction between CH^+ and H_2CO.

$$\begin{aligned}
CH^+ + H_2CO \rightarrow\ & CH_3^+ + CO && (30\%) \\
& H_3CO^+ + C && (30\%) \\
& HCO^+ + CH_2 && (30\%) \\
& H_2C_2O^+ + H && (10\%)
\end{aligned}$$

4.9 Neutralisation processes in the interstellar medium

As noted in Section 4.3, most of the molecules observed in the interstellar medium are neutral, while the products of the ion–molecule reactions discussed in the previous section are ionic. There are a number of pathways by which an ion may be transformed into a neutral molecule within the interstellar medium. We have already covered one of these. In electron–ion dissociative recombination (Section 4.7.2), an ion and electron combine to form a neutral, which then fragments into two or more neutral products. Similarly, a negative ion and a positive ion may recombine to form a neutral complex, which dissociates into neutral products. Some examples of such processes are given below.

$$CO^+ + e^- \to C + O \qquad \text{electron–ion dissociative recombination}$$

$$HCCCNH^+ + e^- \to HCCCN + H \text{ or } HCCNC + H \qquad \text{electron–ion dissociative recombination}$$

$$O^- + O_2^+ \to O + O_2 \qquad \text{ion–ion dissociative recombination}$$

4.10 Summary

We have considered the various chemical environments within interstellar and intergalactic space. We have also explored the consequences of the very low temperatures and pressures within the interstellar medium, both for the types of chemistry that can occur in these regions and for the types of molecules that are stable under these conditions. In order to gain a complete understanding of the chemistry of interstellar space, observational data on the identities and quantities of various molecular species present in interstellar gas clouds must be complemented by data from theoretical and experimental studies carried out on Earth. Over the next two chapters, we will consider the wide range of techniques that have been brought to bear on the challenges of identifying interstellar molecular species and on improving our understanding of their chemistry.

4.11 Questions

4.11.1 Essay-style questions

Q4.1 Many of the molecules formed in interstellar gas clouds have never been observed on Earth. Explain why chemical synthesis in the interstellar medium proceeds very differently from chemical synthesis on Earth.

Q4.2 Outline the primary mechanisms by which ions may be formed in the interstellar medium.

Q4.3 A number of molecules are present in the interstellar medium at number densities that are too high to be explained by gas-phase synthetic pathways. How can their abundance be explained?

4.11.2 Problems

P4.1 *Collision frequencies in the interstellar medium*

Calculate the collision frequency for hydrogen molecules (collision cross-section 0.27 nm^2) in

(a) A sample of H_2 gas at atmospheric pressure.
(b) A giant molecular cloud with $n = 10^5 \text{ cm}^{-3}$ and $T = 40 \text{ K}$.
(c) The diffuse interstellar medium, with $n = 10 \text{ cm}^{-3}$ and $T = 10 \text{ K}$.

[Note that for collisions between particles of the same type, a factor of $1/2$ is required in the expression for the collision frequency to avoid double counting of collisions.]

P4.2 *Ionisation processes in the interstellar medium*

Cosmic rays are highly energetic charged particles — mostly protons, but also some electrons, alpha particles, and heavier nuclei. Collisions of cosmic rays with neutral atoms and molecules provide the primary ionisation mechanism within the interstellar medium.

In a diffuse molecular cloud, the chemistry of the H_3^+ ion is well described by the following sequence of reactions:

$$H_2 + \text{cr} \rightarrow H_2^+ + e^- + \text{cr}' \tag{1}$$

$$H_2^+ + H_2 \rightarrow H_3^+ + H \quad k_2 = 2.08 \times 10^{-9} \text{ cm}^3 \text{ s}^{-1} \tag{2}$$

$$H_3^+ + e^- \rightarrow H_2 + H \tag{3}$$

$$H_3^+ + e^- \rightarrow H + H + H \quad k_3 + k_4 = 6.7 \times 10^{-8} (T/300)^{-0.51} \text{ cm}^3 \text{ s}^{-1} \tag{4}$$

(a) Identify the processes occurring in reactions (1)–(4).
(b) Comment on the temperature dependence of the quantity $(k_3 + k_4)$.

(c) Write down expressions for the rate of change of H_3^+ and H_2^+ number densities in terms of the rate constants k_1 to k_4 and the number densities $n(H_2)$, $n(H_2^+)$, $n(H_3^+)$, $n(\text{cr})$, and $n(e^-)$.

(d) Explain whether the steady-state approximation can be applied to the number densities of H_3^+ and H_2^+ under the typical conditions encountered in a diffuse molecular cloud.

(e) The rate of ionisation α for H_2 is often defined using $\alpha n(H_2) = k_1 n(H_2) n(\text{cr})$. Use this, together with your expression from (b), to show that the ionisation rate can be written

$$\alpha = (k_3 + k_4) x_e n_H \frac{n(H_3^+)}{n(H_2)}$$

where x_e is the electron fraction, defined as $x_e = n(e)/n_H$ and n_H is the total number density of hydrogen atoms, defined as $n_H = n(H) + 2n(H_2)$.

(f) Cosmic ray ionisation rates in diffuse molecular clouds are typically around $3 \times 10^{-16}\,\text{s}^{-1}$, hydrogen densities, n_H, are typically around $100\,\text{cm}^{-3}$, and the electron fraction is around 1.14×10^{-4}. The average temperature in a diffuse molecular cloud is $\sim 70\,\text{K}$. Use this information, together with the rate constant data given in the reaction scheme above, to determine the ratio $n(H_3^+)/n(H_2)$ in a diffuse molecular cloud.

(g) H_3^+ is responsible for most of the proton-transfer chemistry occurring in the interstellar medium. Using the reaction of H_3^+ with water as an example, explain why these reactions occur with such high efficiency.

P4.3 *Synthesis of HF in the interstellar medium*
Atomic fluorine reacts exothermically with H_2 and H_2O as follows, with rate constants as given below.

$$F + H_2 \rightarrow HF + H \quad k_1 = 1.4 \times 10^{-10} \exp(-500\,\text{K}/T)\,\text{cm}^3\,\text{s}^{-1}$$

$$F + H_2O \rightarrow HF + OH \quad k_2 = 1.6 \times 10^{-11}\,\text{cm}^3\,\text{s}^{-1}$$

where T is the temperature in Kelvin. Using your knowledge of the temperature and chemical composition of an interstellar gas cloud, determine which reaction is primarily responsible for formation of HF in interstellar space, explaining your reasoning.

P4.4 *Hydrogen chemistry in an interstellar gas cloud: Steady-state approximation*

A simple model reaction scheme for the hydrogen chemistry occurring within an interstellar gas cloud is given below (the rate constant for each step is given after the relevant reaction equation).

$$H + H \xrightarrow{\text{dust grain}} H_2 \qquad k_1$$
$$H_2 \xrightarrow{h\nu \text{ or cr}} H_2^+ + e^- \qquad k_2$$
$$H_2^+ + H_2 \longrightarrow H_3^+ + H \qquad k_3$$
$$H_3^+ + CO \longrightarrow HCO^+ + H_2 \qquad k_4$$

(a) Explain the chemistry occurring in each step in the reaction sequence.

(b) Apply the steady-state approximation to $[H_2^+]$ and $[H_3^+]$ to obtain an expression for the overall rate of formation of HCO^+.

(c) The steady-state concentration of H_3^+ is 2×10^4 cm^{-3}. The astronomical observation for the column density (the density of H_3^+ along the line of sight) is 3×10^{24} cm^{-2}. Calculate the diameter of the molecular cloud in (i) metres, and (ii) light years.

P4.5 *More hydrogen chemistry in an interstellar gas cloud: simple collision theory*

H_2 can be ionised by a collision with a cosmic ray to form H_2^+, which reacts rapidly with neutral H_2 to form the H_3^+ ion.

$$H_2^+ + H_2 \to H_3^+ + H \qquad (1)$$

According to simple collision frequency, the rate constant of a bimolecular reaction is given by

$$k = \sigma_c \left(\frac{8k_B T}{\pi \mu}\right) \exp(-E_a/k_B T)$$

(a) Define the symbols appearing in the above equation.

(b) Assuming a molecular diameter for both H_2 and H_2^+ of 1.9×10^{-10} m, calculate the collision cross-section.

(c) Use simple collision theory to calculate the rate constant for reaction (1) at a temperature of 30 K.

(d) The rate constant for reaction (1) has been determined from astronomical data to be 2.08×10^{-9} cm^3 s^{-1}. Compare this with your result from (c), above, and suggest reasons to explain any observed discrepancy between the measured value and that calculated from simple collision theory.

Chapter 5

Laboratory-Based Astrochemistry: Theory

5.1 Laboratory-based astrochemistry

While astronomical observations provide raw data on the physical conditions in various regions of space and on the atomic and molecular species present, interpreting such data is only possible as a result of laboratory-based studies aimed at understanding a wide range of basic atomic and molecular processes. For example, molecular identification via the assignment of spectral lines recorded using a telescope can only be achieved through comparison either with spectra recorded on Earth (if the molecule of interest is sufficiently stable and can be generated in sufficiently high quantities to detect) or with spectra simulated on the basis of electronic structure calculations. Kinetic models of interstellar chemistry require knowledge of rate constants for individual chemical steps, which again must either be measured in a laboratory or modelled theoretically. Understanding the complex chemistry occurring on the surface of interstellar dust grains requires the development of suitable laboratory-based analogues and methods for studying them. In many cases, the much higher number densities involved in laboratory experiments, combined with the extremely high detection sensitivities achievable for a wide range of molecules of interest, mean that it is often possible to simulate processes that occur over millions of years in space within a few minutes in the laboratory.

Laboratory-based astrochemistry is a growing field, and an ideal playground for physical chemists, chemical physicists, theoretical chemists, and those working in a number of other related fields. In the following, we will outline a few of the current challenges in astrochemistry that laboratory-based studies can help to address, before looking at some of the available theoretical and experimental methods.

5.2 The grand challenge: Chemical modelling of giant molecular clouds

Perhaps the ultimate goal within the astrochemistry community is to develop a complete chemical model of an interstellar gas cloud. Amongst other things, such a model would allow the age of a molecular cloud to be estimated based on measurements of its chemical composition. The techniques used to model molecular clouds are very similar to those developed for modelling chemical processes in the Earth's atmosphere. However, there are many more unknowns when modelling interstellar gas clouds than in atmospheric modelling. Many of the rate constants, particularly for reactions occurring on dust grains, are unknown, and measurements on interstellar gas clouds to establish parameters such as temperature and number density cannot be carried out directly in the same way as they can in the Earth's atmosphere. Nonetheless, the process of setting up and solving a kinetic model follows the same general principles as for an atmospheric model (see Appendix A for an overview of kinetic modelling), and similar parameters need to be quantified:

(i) Data on the chemical composition of the cloud is taken from experimental observations. Ideally, accurate number densities are required for all chemical species within the cloud, including electrons.
(ii) The physical conditions within the cloud are needed, including temperature, number density, electron temperature, radiation field, and extinction coefficient (used to estimate the dust composition). These conditions will not generally be constant throughout the cloud. For example, the density will fluctuate due to stellar winds, stellar explosions such as supernovae, and random perturbations, while the temperature is determined by the balance of heat input from radiation and exothermic chemical reactions, and heat losses, primarily radiative losses accompanying relaxation of atomic and molecular species from excited states. The radiation field is determined by the proximity to nearby stars and extinction by interstellar dust.

(iii) Transport processes must be considered. These include diffusion and collisions, as well as more exotic phenomena, such as magnetic turbulence or supernova shock waves transiting the cloud.
(iv) An estimate of the radiation field emanating from newly forming stars within the cloud is important in order to account for photochemical processes occurring in these regions.
(v) Accurate reaction rates for all chemical processes occurring in the cloud are needed. Often these are not known, and must be estimated or modelled.
(vi) The reactions to be included in the model must be decided upon, along with the target species that will be compared with available experimental data.

Once all of the required parameters have been quantified (of which more later), the relevant rate equations can be constructed for the chemical reactions of interest (see Appendix A). The rate equations often comprise a complex set of coupled differential equations, which cannot be solved analytically, but can be propagated numerically in time to predict the concentrations of the target species. Some parameters, for example the chemical composition of the cloud and sometimes the temperature in different regions, can be determined from astronomical observation data (see Chapter 1) in combination with data from quantum chemistry calculations. Other parameters, such as rate constants, must be determined in laboratory-based studies, either through experimental measurement or theoretical modelling. To give the reader some sense of the complexity of such models, Figure 5.1 shows the reactions involved in a simplified kinetic model of the chemistry of H_2O in star-forming regions.

In addition to the 'grand challenge' of developing a complete chemical model of an interstellar gas cloud, there are also a number of more specific problems in astrochemistry that are currently attracting considerable attention from the research community. We consider two such examples in the following.

5.2.1 The search for biological molecules

There is enormous interest in detecting biological molecules in interstellar space, as their presence would represent a significant step in the search for extraterrestrial life, as well as potentially providing insight into the

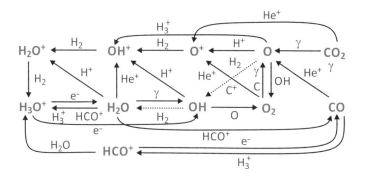

Fig. 5.1 Kinetic model of H_2O chemistry in star-forming environments. *Source*: Adapted from P. Stäuber, J. K. Jorgensen, E. F. van Dishoek, S. D. Doty, and A. O. Benz, Water destruction by X-rays in young stellar objects, *Astron. Astrophys.*, **453**(2), 555 (2006).

origins of life on Earth. As noted in Section 4.8.3.3, the presence of fairly large and complex organic molecules in interstellar space, formed through a combination of ion–molecule reactions and dust-grain-catalysed chemistry, is well established. Laboratory-based studies appear to indicate that there are feasible mechanisms for generating simple amino acids under the conditions known to be present within interstellar gas clouds, and that glycine (the simplest amino acid) and perhaps other biogenic molecules should be abundant enough to be detected. However, to date there has only been one reported detection of glycine,[1] and this has been contested rather than confirmed.[2]

5.2.2 The diffuse interstellar bands (DIBs)

The set of spectroscopic absorption features known as the diffuse interstellar bands represents perhaps the longest-standing mystery in astronomical spectroscopy. The first of these bands was reported[3] in 1922, and observed bands now number more than 400, ranging across the ultraviolet, visible,

---

[1] Y-J. Kuan, S. B. Charnley, H-C. Huang, W-OL Tseng, and Z. Kisiel, Interstellar glycine, *Astrophys. J.*, **593**, 848–867 (2003).
[2] L. E. Snyder, F. J. Lovas, J. M. Hollis, D. N. Friedel, P. R. Jewell *et al.*, A rigorous attempt to verify interstellar glycine, *Astrophys. J.*, **619**, 914–930 (2005).
[3] M. L. Heger, Further study of the sodium lines in class B stars; the spectra of certain class B stars in the regions 5630–6680 Å and 3280–3380 Å; Note on the spectrum of g Cassiopeiae between 5860 Åand 6600 Å, *Lick Observatory Bull.*, **337**, 141–148 (1922).

and infrared regions of the electromagnetic spectrum. Despite general agreement that the absorptions can probably be attributed mostly to large organic molecules in the interstellar medium, there have been no definitive assignments to date. Polycyclic aromatic hydrocarbons (PAHs) (see Section 4.8.3.3), carbon chains, and fullerenes have all been implicated as possible candidates for diffuse interstellar band absorptions. However, the spectroscopy of these species is difficult to study in the laboratory and challenging to model accurately via electronic structure calculations. It is known from solid-state studies that PAH cations comprising 30 or more carbon atoms absorb strongly in the visible, while the corresponding neutrals absorb in the UV, but such large species are very difficult to generate and maintain in the gas phase, a requirement if high-resolution spectroscopic studies are to be performed.

While the infrared emission bands arising from vibrational transitions of PAHs are fairly similar, the electronic spectra are known to be highly characteristic for each molecule. If the problems outlined above are overcome and laboratory-based spectra for large PAHs become available, then such spectra are very likely to allow specific molecules to be identified within the interstellar medium.

5.3   Theoretical astrochemistry I: Spectroscopic data

Ab initio electronic structure calculations are extremely useful in interpreting astrochemical data. Many molecules present in the interstellar medium are extremely difficult to synthesise on Earth, and may be far too reactive under terrestrial conditions to study spectroscopically. In these cases, theoretical data become vital for identifying molecules detected in space and providing information on their structure, spectra, and other properties, including their reactivity. A detailed account of electronic structure methods is beyond the scope of this book. However, in the following we aim to provide a general overview of the theoretical basis of some of these methods, and the ways in which they can be used to help interpret astrochemical data.

All *ab initio* electronic structure calculations are based on numerical solution of the (time-independent) Schrödinger equation for the molecule of interest to find the molecular energy levels and wavefunctions. There are numerous different approaches, but for small to medium-sized molecules of the type found in interstellar gas clouds, most of the more accurate methods are based on the *variation principle*. The variation principle is a fundamental tenet of quantum mechanics which states that any 'trial', or approximate, wavefunction must have a higher energy than the true wavefunction of the system. This is essentially just another way of stating that nature will in general always find the lowest-energy state for a system, and that any guess at that state must therefore either overestimate, or at best equal, this minimum energy. A natural consequence of the variation principle is that if a trial wavefunction has been proposed that contains a number of adjustable parameters, the optimum values for the parameters can be found by minimising the energy of the wavefunction with respect to each parameter. Electronic structure software packages such as Gaussian, [4] GAMESS,[5] MolPro,[6] Spartan,[7] and others construct the wavefunction as a basis set expansion. The approach is similar in spirit to the 'linear combination of atomic orbitals' (LCAOs), approach to constructing molecular orbitals, which will be familiar to many chemists. However, the available basis sets are rather more flexible than the limited set of atomic orbitals usually employed in the LCAO approach, allowing much more accurate wavefunctions to be constructed.

Once an appropriate basis set of functions ($\phi_1, \phi_2, \phi_3, \ldots, \phi_N$) has been chosen, we can write the wavefunctions ψ_i for each state i of the system in the form:

$$\psi_i = c_{1i}\phi_1 + c_{2i}\phi_2 + c_{3i}\phi_3 + \cdots + c_{Ni}\phi_N \tag{5.1}$$

where c_{1i}, c_{2i}, etc. are expansion coefficients denoting the contribution of basis function ϕ_1, ϕ_2, etc. to state i. The energies of the various states are determined by applying the Hamiltonian (energy) operator to the wavefunctions, and are functions of the expansion coefficients, i.e.

$$\hat{H}\psi_i = E_i\psi_i = f_i(c_{1i}, c_{2i}, \ldots, c_{Ni})\psi_i \tag{5.2}$$

Minimising these energy expressions with respect to each coefficient, i.e. setting $dE_i/dc_1 = dE_i/dc_2 = \cdots = 0$, leads to a set of simultaneous equations. These may be solved using standard matrix methods to

---

[4] Gaussian09, Revision D.01, M. J. Frisch et al., Gaussian, Inc, Wallingford CT, 2009, see http://www.gaussian.com.
[5] General Atomic and Molecular Electronic Structure System, M. W. Schmidt, et al., J. Comput. Chem., **14**, 1347–1363 (1993), see http://www.msg.ameslab.gov/gamess/.
[6] MolPro, version 2012.1, a package of ab initio programs, H.-J. Werner, P. J. Knowles, G. Knizia, F. R. Manby, M. Schültz, and others, 2012, see http://www.molpro.net.
[7] Spartan'10, Wavefunction, Inc, Irvine, CA, see http://www.wavefun.com.

determine both the energies of the various states and the optimised expansion coefficients, and therefore the wavefunctions. For keen mathematicians, more details of this process can be found in Appendix B. The energies of the molecular states returned from these calculations are the electrostatic potential energies that arise from interactions between the constituent electrons and nuclei within the molecule. These will turn out to be very important both in determining the energies of various types of spectroscopic transition within the molecule, and in determining reactivity.

Electronic structure calculations can either be carried out at a fixed, user-specified molecular geometry, or alternatively the geometry can be optimised as part of the calculation at the same time as the basis set parameters to give the lowest-energy molecular structure. There are a number of molecular properties that can be calculated for comparison with astrochemical data, as outlined in the following sections.

5.3.1 Rotational transition frequencies

The general steps involved in calculating a rotational or microwave spectrum are shown in Figure 5.2. The molecular geometry may be used to determine the moments of inertia of the molecule about the three Cartesian axes, which in turn allows calculation of the spectroscopic rotational constants and therefore the rotational energy levels.

Molecular structure ⇒ Moments of inertia ⇒ Rotational constants ⇒ Rotational energy levels ⇒ Transition energies

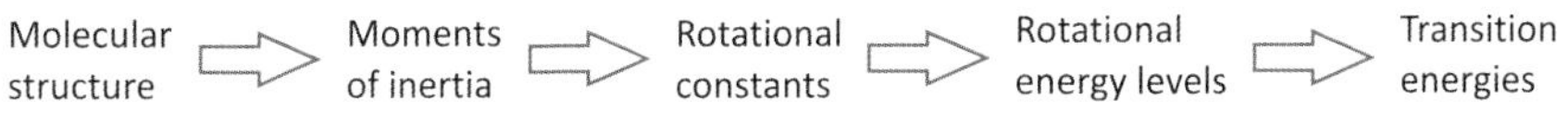

Fig. 5.2 The sequence of steps involved in calculating a rotational absorption or emission spectrum. See text for details.

We will illustrate this process using the simple example of a diatomic molecule AX. The electronic structure calculations return the equilibrium bond length, r_e. The moment of inertia I is then given by

$$I = \sum_i m_i r_i^2 = m_A r_A^2 + m_X r_X^2 \tag{5.3}$$

where m_A and m_X are the masses of A and X, and r_A and r_X are their distances from the axis of rotation, which passes through the centre of mass of the molecule. For a diatomic molecule, it is fairly straightforward (and a good exercise for the reader) to show that this can also be written

$$I = \mu r_e^2 \tag{5.4}$$

where $\mu = m_A m_X/(m_A + m_X)$ is the reduced mass of A and X, and r_e is the equilibrium bond length as before. The rotational constant, B_e (in units of cm^{-1}) is then given by

$$B_e = \frac{h}{8\pi^2 c I} \qquad (5.5)$$

where h is Planck's constant and c is the speed of light (in units of cm s^{-1}). Once the rotational constant is known, the rotational energy levels can be calculated from

$$E_J = B_e J(J+1) - D_e J^2(J+1)^2 \qquad (5.6)$$

where J is the rotational quantum number. The quantity D_e (units of cm^{-1}) is the centrifugal distortion coefficient, and accounts for the fact that the bond stretches slightly as the molecule rotates. Not unsurprisingly, the amount the bond stretches on rotation is related to the force constant k of the bond (i.e. how easy or difficult the bond is to stretch or compress), with the result that D_e can be written in terms of either the force constant or the vibrational wavenumber (in cm^{-1}), ω_e,

$$D_e = \frac{h^3}{32\pi^4 I^2 r_e^2 k c} = \frac{4B_e^3}{\omega_e^2} \qquad (5.7)$$

We will see shortly how the vibrational frequency can be determined from *ab initio* data.

Given the rotational energy levels, it is straightforward to determine the microwave transition frequencies simply by applying the selection rule $\Delta J = \pm 1$, i.e. the rotational quantum number may increase or decrease by one unit on absorption or emission of a microwave photon. Note that in order to possess a microwave (pure rotational) spectrum, the molecule must also satisfy the gross selection rule that it has a nonzero dipole moment.

A similar set of steps can be followed to calculate the rotational transition frequencies of a polyatomic molecule, only in this case (assuming the molecule is nonlinear) the molecule will have three different moments of inertia, I_x, I_y, I_z, corresponding to rotation about the three Cartesian axes, and therefore three different rotational constants, denoted A, B, and C. Polyatomics are generally classified as either symmetric tops, characterised by two of their three moments of inertia being identical, or asymmetric tops, in which case all three moments of inertia are different. In addition to the quantum numbers J and M_J, characterising the magnitude of the rotational angular momentum and its projection onto a laboratory axis, symmetric and asymmetric tops have a third quantum number, K, denoting the projection of the angular momentum vector $\mathbf{J}$ onto the principal axis of the molecule. The energy levels of a symmetric top, neglecting contributions from centrifugal distortion, are given by

$$E_{JK} = BJ(J+1) + (A-B)K^2 \tag{5.8}$$

and the selection rules are $\Delta J = \pm 1$, $\Delta K = 0$. For asymmetric tops it is not possible to obtain a straightforward analytical expression for the rotational energy levels, and for a given molecule these must instead be determined by solving the Schrödinger equation for each J value.

5.3.2 Vibrational transition frequencies

The vibrational motion of a diatomic molecule is determined by the way in which the molecular potential energy, as calculated in the electronic structure calculations, depends upon the distance between the two atoms, i.e. the bond length. As we shall show in the following, by considering derivatives of the potential energy, it is possible to determine the vibrational frequencies of a molecule, and therefore to predict its infrared (vibrational) spectrum.

The vibrational frequencies can be calculated by approximating the bottom of the vibrational potential to that of a harmonic oscillator. In reality, vibrational potentials are Morse potentials of the type shown in Figure 5.3, but a harmonic potential is a good approximation at the

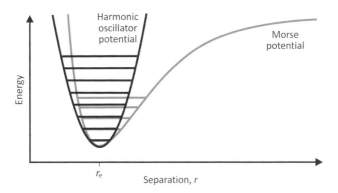

Fig. 5.3 The vibrational potential energy curve of a diatomic molecule is a Morse potential, but is often approximated by a harmonic oscillator potential.

minimum, corresponding to the equilibrium bond length or angle. It should be noted, however, that the harmonic approximation fails badly far from the equilibrium position, and does not allow for bond dissociation when molecules are highly vibrationally excited.

Continuing with the example of a diatomic molecule, the harmonic oscillator potential $V(r)$ for the stretching vibration is given by

$$V(r) = \frac{1}{2}k(r - r_e)^2 \tag{5.9}$$

where r is the bond length at a given point in the vibrational motion, r_e is the equilibrium bond length, and k is the force constant of the bond. For a harmonic oscillator, the force constant is related to the vibrational frequency ν and the reduced mass μ of the molecule by

$$\nu = \frac{1}{2\pi}\left(\frac{k}{\mu}\right)^{1/2} \quad \text{or} \quad \omega_e = \frac{1}{2\pi c}\left(\frac{k}{\mu}\right)^{1/2} \tag{5.10}$$

where ω_e is the vibrational wavenumber, usually expressed in units of cm^{-1}. Inspection of Equation (5.9) reveals that the force constant can be determined in a straightforward way by taking the second derivative of the potential.

$$\frac{\mathrm{d}^2 V(r)}{\mathrm{d}r^2} = k \tag{5.11}$$

Substituting this result into Equation (5.10) yields

$$\omega_e = \frac{1}{2\pi c}\left(\frac{1}{\mu}\frac{\mathrm{d}^2 V(r)}{\mathrm{d}r^2}\right)^{1/2} \tag{5.12}$$

The vibrational energy levels of the diatomic molecule are then simply the energy levels of a quantum mechanical harmonic oscillator, namely

$$E_v = hc(v + 1/2)\omega_e \tag{5.13}$$

where v is the vibrational quantum number. As noted, the above treatment approximates the true vibrational potential to that of a harmonic oscillator (parabolic) potential. More sophisticated treatments include anharmonic effects by taking the above expression for the energy levels as the first term in a Taylor series expansion describing the true energies, with the higher order terms constituting a series of *anharmonic corrections* to the harmonic oscillator energy levels

$$E_v = (v + 1/2)\omega_e - (v + 1/2)^2 x_e\omega_e + (v + 1/2)^3 y_e\omega_e + \cdots \tag{5.14}$$

where the quantities $x_e\omega_e$ and $y_e\omega_e$ are called anharmonicity constants. These, along with the vibrational frequency ω_e are tabulated in numerous spectroscopic databases for a wide variety of molecules.[8]

In order to absorb or emit infrared radiation, the molecule must obey the gross selection rule that the vibrational motion is accompanied by a corresponding change in the molecular dipole moment. Note that because the molecule will virtually always be rotating as well as vibrating, each vibrational level contains a manifold of rotational levels, such that transitions will be from some initial *rovibrational* state (v_i, J_i) to some final state (v_f, J_f). The allowed transitions must satisfy both the vibrational selection rule $\Delta v = \pm 1, \pm 2, \pm 3, \ldots$[9] and the rotational selection rule $\Delta J = 0, \pm 1$. Transition energies are then simply calculated as the difference between the energies of the final and initial rovibrational levels.

Diatomic molecules have only one bond, and therefore only one possible (stretching) vibration. The vibrational structure of polyatomic molecules, which can undergo a variety of different stretching and bending vibrations, is considerably more complex. The number of different vibrational modes or vibrational degrees of freedom of an N-atom molecule is simple to determine based on the $3N$ total degrees of freedom (i.e. three independent directions of motion for each atom). Three of these $3N$ degrees of freedom correspond to translational of the whole molecule, and two (for linear molecules, with two independent axes of rotation) or three (for nonlinear molecules with three axes of rotation) correspond to rotational degrees of freedom of the molecule. Note that the moment of inertia associated with a linear molecule 'spinning' on its molecular axis is zero, so there is no kinetic energy associated with this motion and rotation about this axis is not a valid degree of freedom. The remaining degrees of freedom once translation and rotation have been accounted for must be vibrational, yielding $3N - 5$ vibrational modes if the molecule is linear and $3N - 6$ if it is nonlinear. Determining the molecular motions and vibrational frequencies associated with the various modes from first principles is relatively complex, and requires some knowledge of group theory. However, both of these properties are routinely provided as output from electronic structure calculation packages if vibrational frequency calculations have been requested.

---

[8] See, for example, the NIST chemistry webbook at http://webbook.nist.gov.
[9] For transitions occurring from $v = 0$, the $\Delta v = 1$ transition is referred to as the *fundamental*, with the $\Delta v = 2, 3, 4$, etc. transitions referred to as the first, second, and third *overtone*, and so on. Transitions occurring from $v = 1$ and higher levels are known as *hot bands*. Their intensities increase with temperature as the population of excited states in a sample increases.

5.3.3 Electronic transition frequencies

The electronic structure calculations described in Section 5.3 yield a set of molecular orbitals for the molecular of interest, together with the energy of the ground electronic state, i.e. the state corresponding to the most stable configuration of electrons within the calculated orbitals. Using more sophisticated electronic structure methods, it is also possible to calculate the energies of excited electronic states, and hence to calculate the frequencies of electronic transitions within the molecule. For a diatomic molecule, optically allowed transitions obey the following electronic selection rules:

$\Delta S = 0$ Spin selection rule
$\Delta \Lambda = 0, \pm 1$ Orbital angular momentum selection rule
$u \leftrightarrow g$ Symmetry selection rules
$+ \leftrightarrow +$
$- \leftrightarrow -$

where S is the total spin angular momentum quantum number, and Λ is the quantum number quantifying the projection of the total orbital angular momentum **L** onto the inter-nuclear axis. The letters g and u (from the German *gerade* and *ungerade*) denote the inversion symmetry of the state in molecules possessing a centre of inversion, while '+' and '−' denote the symmetry with respect to reflection in a plane containing the inter-nuclear axis.

5.3.4 Transition intensities

The transition intensity associated with a transition between an initial state i and a final state f is determined by the product of the transition probability P_if and the population difference Δp between the lower and upper states.[10] The transition probability depends on the initial and final state

---

[10] The dependence of the transition intensity on the population difference between the initial and final states arises from the fact that the processes of stimulated absorption (in which an incident photon of the appropriate energy induces excitation of a molecule from a lower to an upper state) and stimulated emission (in which an incident photon induces de-excitation of a molecule from an upper state to a lower state) have equal probabilities. The rates of stimulated absorption and stimulated emission at a frequency ν are equal to $B_{12}\rho(\nu)N_1$ and $B_{21}\rho(\nu)N_2$, respectively. Here, $B_{12} = B_{21}$ is known as the *Einstein B coefficient*, $\rho(\nu)$ is the density of photons of frequency ν, and N_1 and N_2 are the number of molecules in the lower and upper states, respectively. The net rate of absorption (i.e. the fraction not 'cancelled out' by stimulated emission) is therefore $B_{12}\rho(\nu)(N_1-N_2)$, and is proportional to the population difference N_1-N_2 between the two states.

wavefunctions ψ_i and ψ_f and on the dipole operator, $\hat{\mathbf{d}}$.
$$P_{if} = |\langle\psi_f|\hat{\mathbf{d}}|\psi_i\rangle|^2 \tag{5.15}$$
The population of a given state j at temperature T is determined by the Boltzmann distribution (derived in any standard statistical mechanics text),
$$p_j = \frac{g_j \exp(-\epsilon_j/k_B T)}{q} \tag{5.16}$$
where g_j is the degeneracy and ϵ_j the energy of the state, k_B is Boltzmann's constant, and q is the molecular partition function, a measure of the total number of thermally accessible states.
$$q = \sum_j g_j \exp(-\epsilon_j/k_B T) \tag{5.17}$$

Without going into detailed explanations of the origins of Equations (5.15)–(5.17), which can be found in many texts on Spectroscopy and Statistical Mechanics, the key point to appreciate is that, given knowledge of the molecular wavefunctions and energy levels, transition intensities can be predicted. Note that *the Boltzmann distribution only applies for a thermal distribution of molecules*, which is not always the case in the interstellar medium.

5.4 Theoretical astrochemistry II: Kinetic and dynamical data

In the previous section, we showed that data from electronic structure calculations carried out at the equilibrium geometry of a molecule of interest can be used to obtain spectroscopic data for comparison with observations. Predicting the outcome of a collision between two molecules — for example, whether the collision leads to elastic, inelastic, or reactive scattering (these terms will be defined shortly), and the mechanism and rate of any reactions that occur — is a rather more complicated proposition. Rather than simply requiring knowledge of the molecular energy levels at a single fixed geometry, a good understanding of the *complete energetic pathway* linking reactants and products is required. In addition, we need to introduce some basic concepts from the field of collision physics. The following sections will provide an introduction both to the language of collision dynamics and to the important concept of a potential energy surface (PES) for a chemical reaction. Armed with an understanding of these concepts, we will then be ready to explore various approaches to modelling the kinetics and dynamics of reactions occurring in the interstellar medium.

5.4.1 Types of collision

Collisions may be categorised into three types:

(1) Elastic collisions are collisions in which kinetic energy is conserved. The collision partners have the same total translational kinetic energy after the collision as they did before the collision, and no energy is transferred into internal degrees of freedom (rotation, vibration, etc.). Note that the kinetic energy of each individual collision partner may change, but the total kinetic energy remains constant.
(2) Inelastic collisions are collisions in which kinetic energy is not conserved, and energy is converted between different forms, e.g. translational to rotational or vibrational. Reactive collisions are a subset of inelastic collisions, but the term 'inelastic' is usually used to describe collisions in which translational kinetic energy is transformed into internal excitation of one or both of the collision partners (or vice versa).
(3) A reactive collision is a collision leading to a chemical reaction, i.e. a collision in which chemical bonds are made or broken, so that the species leaving the collision region are chemically distinct from those that entered. The collision energy must be high enough to overcome any activation barrier associated with reaction. As noted above, reactive collisions are a special case of inelastic collisions.

Reactive collisions are the processes by which chemical change occurs, while inelastic collisions are often the means by which molecules gain enough energy to overcome activation barriers so that subsequent collisions may lead to reaction.

All collisions must obey certain physical laws, such as conserving energy and both linear and angular momentum. In addition, there are a number of concepts associated with collisions that will probably not be familiar to many readers, and which will now be introduced.

5.4.2 Relative velocity

Rather than considering the individual velocities,[11] $\mathbf{v_1}$ and $\mathbf{v_2}$, of the two collision partners, it is often more useful to consider their relative velocity. The relative velocity is simply the difference between the individual

---

[11] Note that velocity is a vector quantity. In the following, vectors are indicated by bold letters, and their (scalar) magnitude by the corresponding non-bold letter.

velocities, and corresponds to the 'apparent' velocity of the second particle relative to an observer travelling along with the first particle (or vice versa)

$$\mathbf{v_{rel}} = \mathbf{v_1} - \mathbf{v_2} \tag{5.18}$$

The usefulness of the relative velocity when considering the likely outcome(s) of a collision may be illustrated with a simple everyday analogy. Imagine you are sitting on a moving train and measuring the apparent velocity of a second train in front of you. If the second train is travelling in the same direction as your train (i.e. away from your train), it will appear to have a lower velocity than its true velocity, while if it is travelling in the opposite direction (i.e. towards your train) it will appear to have a higher velocity. As will be demonstrated quantitatively in Section 5.4.3, the relative velocity is very important in reaction dynamics (and in train crashes!), as it determines the collision energy, and therefore the probability of overcoming any energy barrier to reaction. To extend the train analogy above, the collision energy is much lower when the two trains are travelling in the same direction (low relative velocity) than when they suffer a head-on collision (high relative velocity).

5.4.3 Collision energy, total kinetic energy, and conservation of linear momentum

The total kinetic energy of the colliding particles is simply equal to the sum of their individual kinetic energies:

$$K = \frac{1}{2}m_1 v_1^2 + \frac{1}{2}m_2 v_2^2 \tag{5.19}$$

where m_1 and m_2 are the masses of the two particles. Alternatively, the total kinetic energy can be broken down into two contributions, one associated with the velocity of the centre of mass of the collision partners, and one associated with their relative velocity

$$K = \frac{1}{2}M v_{CM}^2 + \frac{1}{2}\mu v_{rel}^2 \tag{5.20}$$

Here, $M = m_1 + m_2$ is the total mass, μ the reduced mass of the two particles, v_{CM} the velocity of the centre of mass of the particles, and v_{rel} their relative velocity. The velocity of the centre of mass is calculated

from
$$\mathbf{v_{CM}} = \frac{m_1\mathbf{v_1} + m_2\mathbf{v_2}}{M} \tag{5.21}$$

The total linear momentum of the two particles must be conserved throughout the collision, and is equal to

$$\mathbf{p} = m_1\mathbf{v_1} + m_2\mathbf{v_2} = M\mathbf{v_{CM}} \tag{5.22}$$

Note that Equations (5.21) and (5.22) are equivalent. Since the momentum of the centre of mass is conserved, the kinetic energy associated with the motion of the centre of mass, $K = \frac{1}{2}Mv_{CM}^2 = p^2/2M$, must also be conserved throughout the collision, and is therefore not available to overcome any energetic barriers to reaction. Only the kinetic energy contribution arising from the relative velocity of the two particles is available for reaction. This energy is known as the collision energy.

$$E_{coll} = \frac{1}{2}\mu v_{rel}^2 \tag{5.23}$$

5.4.4 Conservation of energy and energy available to the products

The total energy must be conserved in any collision. Before the collision, the total energy is the sum of contributions from the kinetic energies of the collision partners (or, equivalently, the kinetic energies associated with their relative motion and the motion of their centre of mass) and any internal energy associated with rotation, vibration, or electronic excitation of the reactants. The reaction itself may consume energy (in the case of an endoergic reaction, with a positive $\Delta_r H$) or release energy (in the case of an exoergic reaction, with a negative $\Delta_r H$), and this must be taken into account when considering the amount of energy available to the products.

$$E_{avail} = E_{coll} + E_{int}(\text{reactants}) - \Delta_r H_0 \tag{5.24}$$

Note that $\Delta_r H_0$ denotes the enthalpy of reaction at a temperature of $0\,\mathrm{K}$ (i.e. for ground-state reactants reacting to form ground-state products). The energy available to the products may be distributed between translational and internal energy of the products.

5.4.5 Impact parameter, b, and opacity function, $P(b)$

The impact parameter quantifies the initial perpendicular separation of the paths of the collision partners (see Figure 5.4). Essentially, this is the

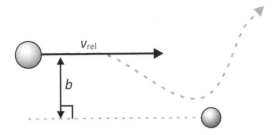

Fig. 5.4 The impact parameter, b, denotes the smallest perpendicular distance between the reactants in the absence of interactions.

distance by which the colliding pair would miss each other if they did not interact in any way, and can be found by extrapolating the initial straight-line trajectories of the particles at large separations to the distance of closest approach.

The probability of reaction as a function of impact parameter, $P(b)$, is called the *opacity function*.

5.4.6 Collision cross-section, σ_c

Many readers will have come across the collision cross-section before, perhaps in the context of the simple collision theory of chemical kinetics (see Equation (4.1) and associated discussion in the previous chapter, for example). The collision cross-section defines the collision probability in terms of an area, and may be thought of as the cross-sectional area that the centres of two particles must lie within if they are to collide. A very simple model can be used to gain a conceptual understanding of the collision cross-section and its relationship to the collision rate. In the kinetic model of gases, it is assumed that there are no interactions between gas particles, and that the particles therefore act like hard spheres or 'billiard balls'. Figure 5.5 shows a few particles within such a gas. Imagine that the the motion of all of the particles has been frozen, apart from that of the darker coloured particle positioned on the left in the side-on view, and in the foreground in the head-on view. Assuming the particle continues through the gas in a straight line, the moving particle will only collide with other particles whose centres lie within the cross-sectional area $\sigma_c = \pi d^2$, where d is the particle diameter. We define the quantity σ_c as the collision cross-section.

The total number of collisions the particle will undergo in a given time period is simply the number of particles whose centres lie within the

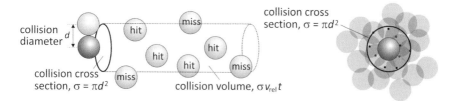

Fig. 5.5 The collision cross-section defines the cross-sectional area within which the centres of two particles must lie if they are to collide.

cylindrical volume swept out by the cross-sectional area in the chosen time period. If the relative velocity of the particles is $v_{\rm rel}$, then the distance travelled (in a straight line) by the moving particle in a time t is $v_{\rm rel}t$, and the volume swept out is $\sigma_c v_{\rm rel} t$. If the number density (number of particles per unit volume) of the stationary particles is n, then the total number of collisions the moving particle undergoes is $\sigma_c v_{\rm rel} t n$, and the number of collisions per unit time is $\sigma_c v_{\rm rel} n$. This relationship still holds if we allow all of the particles to be moving. A simple first-order rate law describing the rate of collisions of the moving particle predicts the collision rate to be $k_c n$, where k_c is the collision rate constant and n is the number density of the particles, as before. Comparing the two expressions reveals the relationship between the rate constant and the collision cross-section, namely $k_c = \sigma_c v_{\rm rel}$. The collision cross-section may therefore be interpreted essentially as a 'microscopic' or 'single collision' version of the collision rate constant.

In the more general case of interacting particles, the action of attractive interactions between particles often results in trajectories with impact parameters larger than the particle diameter, i.e. $b > d$, leading to collisions. In this case, the collision cross-section becomes somewhat larger than the hard-sphere collision cross-section illustrated above.

5.4.7 Reaction cross-section, σ_r

The reaction cross-section σ_r also appears in the context of simple collision theory, but is again a general property of any reactive collision. In the context of simple collision theory, the reaction cross-section is found by multiplying the collision cross-section by a 'steric factor', i.e. $\sigma_r = P\sigma_c$, which accounts for the fact that geometrical requirements mean that in general not all collisions lead to reaction. A somewhat more rigorous treatment than that provided by simple collision theory is required in the present context.

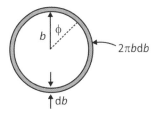

Fig. 5.6 Volume elements involved in the definition of the reaction cross-section.

In the same way as the collision cross-section is a microscopic version of the collision rate constant (see Section 5.4.6), the reaction cross-section is the microscopic or single-collision version of a reaction rate constant, representing the reaction probability as an effective cross-sectional area within which the two reactants must collide in order for reaction to occur. For a reaction between two chemical species that can be modelled as hard spheres, so that the reaction probability is uniform for impact parameters between zero and some maximum value b_{max}, the reaction cross-section is

$$\sigma_{\text{r}} = \pi b_{\text{max}}^2 \tag{5.25}$$

More generally, the reaction cross-section is given by integrating the reaction probability $P_{\text{r}}(b)$ as a function of impact parameter and angle.

$$\sigma_{\text{r}} = \int_0^{2\pi} \int_0^{b_{\text{max}}} P_{\text{r}}(b) b\, \text{d}b\, \text{d}\phi = 2\pi \int_0^{b_{\text{max}}} P_{\text{r}}(b) b\, \text{d}b \tag{5.26}$$

The volume element $2\pi b\, \text{d}b$ is illustrated in Figure 5.6.

5.4.8 The excitation function, $\sigma_{\text{r}}(E_{\text{coll}})$, and the thermal rate constant, $k(T)$

The excitation function quantifies the dependence of the reaction cross-section on collision energy, and is therefore the microscopic equivalent of the temperature dependence of the rate constant. Broadly speaking, there are two different types of excitation function, shown schematically in Figures 5.7(a) and 5.7(b), with the type exhibited by a particular reaction depending on whether or not the reaction has an activation barrier.

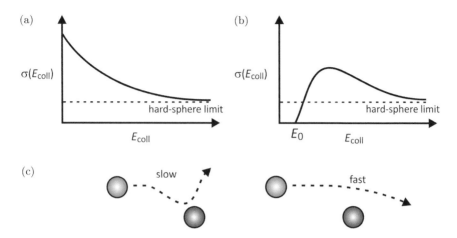

Fig. 5.7 Generic excitation functions for exoergic reactions (a) without an activation barrier and (b) with a barrier; (c) At low relative velocities, attractive forces act over relatively long periods of time to pull particles together, leading to a large reaction cross-section; at high relative velocities, there is little time for attractive interactions to act, and trajectories are deflected only slightly, if at all, leading to small or 'hard-sphere' reaction cross-sections.

5.4.8.1 *Exoergic with no barrier*

This type of excitation function is typical of many exoergic[12] processes, e.g. ion–molecule reactions. The origin of the variation in cross-section with energy is shown schematically in Figure 5.7(c). At low energies there is enough time for long-range attractive interactions (van der Waals interactions, dipole interactions, and so on) to act on the particles and cause significant deviations in their trajectories. If the long-range interactions are strong, even particles with large impact parameters can be pulled together by the attractive interactions and undergo a collision, leading to a large collision cross-section, and therefore a large reaction cross-section. As the collision energy increases, the particles fly past each other faster and their trajectories are deviated by a smaller amount by the attractive interactions. The cross-section therefore reduces, until eventually there is almost

---

[12]When talking about single reactive collisions, the terms *endoergic* and *exoergic* are used in place of *endothermic* and *exothermic*.

no time for any attractive forces to act and the so-called 'hard-sphere limit' is reached. At this point, the collision cross-section is given by $\sigma_c = \pi d^2$, where $d = r_A + r_B$ is the sum of the particle radii, with the reaction cross-section reducing accordingly. For many reactions the hard-sphere limit is very small, and the excitation function essentially tends to zero at very high energies. In Section 5.4.20, we will derive a simple model for the rate constant of an ion–molecule reaction that reproduces these features.

5.4.8.2 *Endoergic or exoergic with a barrier*

When the reaction has a barrier, no reaction occurs until there is enough energy available to surmount the barrier, and the cross-section is therefore zero at collisions energies below this threshold. As the collision energy is increased above the energetic barrier to reaction, the cross-section increases to a maximum over the range of collision energies for which attractive forces can act, displaying an 'Arrhenius-type' energy dependence (see Appendix A), before decreasing again to the hard-sphere limit at high energies, for the same reasons as described above. This type of excitation function is essentially the product of the decreasing 'barrierless' excitation function described in Section 5.4.8.1 and an 'Arrhenius-type' function which rises once the activation barrier is surmounted.

Once the excitation function is known, expressed either as a function of collision energy or relative velocity, the thermal rate constant may be determined. The rate constant for a given relative velocity is simply $k(v) = v\sigma_r(v)$, so integrating over the velocity distribution $P(v)$ of the molecules present in the sample yields the thermal rate constant.

$$k(T) = \int_0^\infty v\sigma_r(v)P(v)dv \qquad (5.27)$$

In most terrestrial applications, the velocity distribution $P(v)$ is simply a thermal Maxwell–Boltzmann distribution. However, this may not be the case in the rarefied conditions of the interstellar medium.

5.4.9 Orbital angular momentum, L, and conservation of angular momentum

In the context of a collision, the orbital angular momentum is a classical angular momentum associated with the relative motion of the collision partners as they approach and collide. It is *not* to be confused with the

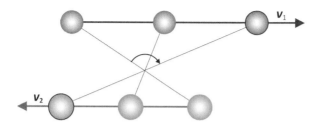

Fig. 5.8 The line of centres of two approaching particles rotates about their centre of mass, giving rise to orbital angular momentum.

quantum mechanical orbital angular momentum of an electron in an atomic orbital. Even for two particles travelling in completely straight lines, there is an associated orbital angular momentum when their relative motion is considered. This is illustrated in Figure 5.8, which plots the line of centres of the two particles at various points in their trajectory. It is clear that even though the individual particles are both travelling in straight lines, the line of centres of the particles rotates about their centre of mass. Only head-on collisions with an impact parameter $b = 0$ have no associated orbital angular momentum.

Mathematically, the orbital angular momentum for a colliding pair of particles is given by

$$\mathbf{L} = \mathbf{r} \times \mathbf{p} \tag{5.28}$$

where $\mathbf{r}$ is the (vector) separation of the particles and $\mathbf{p}$ is their relative linear momentum. We can therefore find the magnitude of $\mathbf{L}$ from

$$|\mathbf{L}| = |\mathbf{r} \times \mathbf{p}| = |\mu \mathbf{r} \times \mathbf{v}_{\text{rel}}| = \mu v_{\text{rel}} r \sin\theta \tag{5.29}$$

where θ is the angle between $\mathbf{r}$ and $\mathbf{v}_{\text{rel}}$. At large separations, $r\sin\theta$ is simply equal to the impact parameter, b, giving

$$|\mathbf{L}| = \mu v_{\text{rel}} b \tag{5.30}$$

Because the total angular momentum (the sum of the orbital angular momentum $\mathbf{L}$, and any rotational angular momentum $\mathbf{J}$, of the collision partners) must be conserved throughout the collision, Equation (5.30) holds right up until the point that the particles collide, assuming that the rotational states of the particles do not change. Note that the relative velocity and impact parameter may change from their initial values as the particles approach, due to the action of attractive or repulsive intermolecular forces.

In the products, the total angular momentum, **L** + **J**, must be the same as before the collision, but angular momentum can be (and often is) exchanged between **L** and **J** during the collision.

5.4.10 The interaction potential and its effect on the collision cross-section

The effect of atomic and molecular interactions on reaction cross-sections was discussed in a qualitative way in Section 5.4.8, in the context of their effect on the excitation function. There we introduced the idea that attractive interactions between two colliding particles can increase the cross-section above that expected for a 'hard-sphere' collision. We will now provide a somewhat more quantitative treatment in the context of collision cross-sections.

The potential energy of interaction between a colliding pair of atoms depends on a single coordinate, namely the separation of the atoms. The simplest model is the hard-sphere potential, in which there is no interaction until the two spheres touch. The maximum impact parameter is $b_{\max} = d = r_1 + r_2$, the sum of the radii, and the collision cross-section is $\sigma_c = \pi b_{\max}^2$, and is not energy dependent. The interaction potential, an example collision trajectory, and the cross-section for such a system are shown in Figure 5.9.

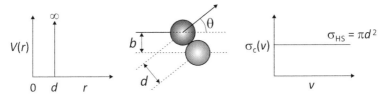

Fig. 5.9 From left to right: the interaction potential, an example trajectory, and the velocity-dependent collision cross-section for a hard-sphere collision.

In contrast, a realistic interatomic potential has a long-range attractive component and a short-range repulsive component. Often, the Lennard-Jones[13] potential is used, which has a repulsive component proportional to $1/r^{12}$ and an attractive component proportional to $1/r^6$.

$$V(r) = \epsilon \left(\left(\frac{r_0}{r}\right)^{12} - 2\left(\frac{r_0}{r}\right)^6 \right) \tag{5.31}$$

---

[13] J. E. Lennard-Jones, On the determination of molecular fields, *Proc. R. Soc. Lond. A*, **106**(738), 463–477 (1924).

Here, ϵ is the depth of the potential well at the minimum, and r_0 is the particle separation at the minimum. The Lennard-Jones potential is shown in Figure 5.10, together with an example collision trajectory and collision cross-section. In contrast to the hard-sphere potential, the impact parameter can now exceed the collision diameter d due to the attractive interaction between the particles. The two particles are attracted to each other as they approach, but are then scattered before actually touching by the repulsive part of the potential.

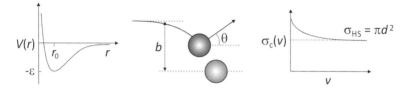

Fig. 5.10 The interaction potential, an example trajectory, and the velocity-dependent collision cross-section for a collision governed by a Lennard-Jones potential.

Landau, Lifshitz, and Schiff[14] proposed an expression to relate the collision cross-section to the Lennard-Jones parameters ϵ and r_0 and to the relative velocity v of the colliding particles.

$$\sigma_c(v) = 8.083 \left(\frac{2\epsilon r_0^6}{\hbar v} \right)^{2/5} \qquad (5.32)$$

Note that the $1/v^{2/5}$ dependence of the collision cross-section in Equation (5.32) yields a function similar to that introduced in the excitation function for reactions without a barrier, discussed in Section 5.4.8.

It has hopefully become clear that the collision cross-section depends on the interactions between the colliding particles, and therefore that measurement of the collision cross-section can provide information about these interactions. In the following sections, we will explore molecular interactions in more detail, and will demonstrate that the interactions between colliding atoms and molecules in fact completely determine every aspect of the collision dynamics.

---

[14]R. B. Bernstein and K. H. Kramer, Comparison and appraisal of approximation formulas for total elastic molecular scattering cross-sections, *J. Chem. Phys.*, **38**(10), 2507–2511 (1963).

5.4.11 Atomic and molecular interactions

The Lennard-Jones potential introduced above provides a simple two-parameter approximation to the true interaction potential between two atoms or molecules. More realistic potentials take into account all of the various types of atomic and molecular interactions that can contribute to the attractive part of the interaction potential. The various contributions can all be written in the form

$$V(r) = -\frac{a}{r^n} \tag{5.33}$$

where a is a constant that depends on electrostatic properties such as the atomic or molecular charge, dipole moments, polarisabilities, etc., and the exponent n depends on the type of interaction. The values of a and n for the various different types of molecular interaction are given in Table 5.1.

5.4.12 The potential energy surface for a polyatomic system

For a collision between two atoms, the interaction potential energy depends on only a single parameter, the inter-nuclear separation. When two *molecules* collide, there are many more degrees of freedom, associated with the positions of each of the individual atoms, all of which may be important in determining the outcome of a collision. Figure 5.11(a) shows the

Table 5.1 Values of a and n for molecular interactions with potential energy $V(r) = -a/r^n$.

Interaction	n	a
Ion–ion	1	$q_1 q_2/(4\pi\epsilon_0)$
Ion–dipole	2	$\mu_1 q_2/(4\pi\epsilon_0)$
Dipole–dipole (aligned dipoles)	3	$\mu_1 \mu_2/(2\pi\epsilon_0)$
Ion–induced dipole	4	$q_1^2 \alpha_2/(8\pi\epsilon_0)$
Dipole–dipole (freely rotating)	6	$2\mu_1^2 \mu_2^2/(3(4\pi\epsilon_0)^2 k_B T)$
Dipole–induced dipole	6	$\mu_1^2 \alpha_2/(4\pi\epsilon_0)$
Induced dipole–induced dipole interaction	6	$3\alpha_1 \alpha_2 I_1 I_2/(2(I_1 + I_2))$

Note: Induced-dipole–induced-dipole interactions are also known as *London* or *dispersion* interactions, and that all of the interactions between closed shell molecules are often referred to collectively as *van der Waals interactions*. In the following, q denotes a charge, μ a dipole moment, α a polarisability volume, and I an ionisation energy, and the subscripts 1 and 2 denote the two interacting molecules. T is temperature, and ϵ_0 and k_B are the permittivity of free space and Boltzmann's constant, respectively.

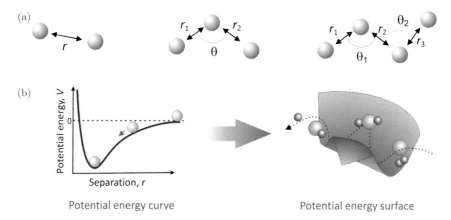

Fig. 5.11 (a) Coordinates needed to define the relative positions of two, three, and four atoms; (b) schematic illustration of a potential energy curve for a diatomic collision, and a potential energy surface for a many-body collision.

way in which the number of degrees of freedom describing the collision increases with the number of atoms in the collision system. Specifying the relative position of two atoms requires a single coordinate; for three atoms we require two interatomic distances and an angle; for four atoms we require three interatomic distances, two bond angles, and a dihedral angle, and so on. Generalising, the total number of coordinates required to describe the positions of N atoms is $3N$. However, the energy of the system depends only on the *relative* coordinates of the atoms. These are not changed by a translation of the whole system or by rotation of the whole system, each of which require 3 of the $3N$ total degrees of freedom, so the total number of coordinates required in general is $3N - 6$. For a linear system, the number of coordinates required is $3N - 5$, since only two degrees of freedom are needed to describe a rotation of the whole system in this case.

In order to describe the potential energy of interaction associated with anything more complicated than a collision between two atoms, it is therefore necessary to move on from the simple idea of a potential energy curve to a potential energy *surface*. The complete potential energy surface quantifies the potential energy of the system as a function of all of the nuclear coordinates of the atoms involved, or alternatively in terms of bond lengths, bond angles, etc., Each point on the surface corresponds to a particular geometrical configuration of atoms, and there may be several potential energy surfaces for a given collision system, corresponding to different possible electronic states of the atoms or molecules involved. The move from a potential energy curve to a potential energy surface is illustrated schematically in Figure 5.11(b).

5.4.13 Construction of the potential energy surface

In order to calculate a single point on the potential energy surface the Schrödinger equation must be solved to determine the energy at the atomic configuration of interest. While it is not possible to solve the Schrödinger equation exactly, as has already been outlined in Section 5.3, there are numerous electronic structure software packages available (e.g. Gaussian, Gamess, MolPro), all of which can solve the equation numerically to a high degree of accuracy. In principle, to construct the complete potential energy surface the Schrödinger equation would have to be solved for every possible atomic configuration. However, in practice, we solve for a reasonable number of configurations and then fit the resulting data points to a suitable function in order to generate a smooth surface. Though this procedure is straightforward in principle, the construction of an accurate potential energy surface for a chemical system is a challenging problem, which quickly becomes intractable as the number of atoms in the system, and therefore the number of degrees of freedom required, increases. Nonetheless, accurate surfaces are available for a considerable number of reactive chemical systems, and comparing the predictions of these surfaces with the results of experimental measurements provides a stringent test both for theory and experiment. Note that the construction of potential energy surface assumes the Born–Oppenheimer approximation, i.e. that the electronic and nuclear motion can be decoupled, so that it is valid to calculate the electronic energy at a series of fixed nuclear geometries.

5.4.14 The potential energy surface and the collision dynamics

Having introduced the concept of a reaction potential energy surface the next step is to explain the link between the potential energy surface and the collision dynamics. In fact, the link is very straightforward. Once the potential energy of a system is known as a function of the relevant coordinates, the force acting on the system along a particular coordinate may be determined from the first derivative or gradient of the potential along that coordinate. For a potential energy function that depends on a single coordinate, r, we have

$$F(r) = -\frac{\mathrm{d}V(r)}{\mathrm{d}r} \qquad (5.34)$$

which for a potential energy function that depends on coordinates r_1, r_2, r_3, etc., generalises to

$$\mathbf{F} = -\nabla V = -\left(\frac{\partial V}{\partial r_1}, \frac{\partial V}{\partial r_2}, \frac{\partial V}{\partial r_3}, \ldots\right) \qquad (5.35)$$

Here, the operator $\nabla = (\frac{\partial}{\partial r_1}, \frac{\partial}{\partial r_2}, \frac{\partial}{\partial r_3}, \ldots)$ is the gradient operator, and the resulting vector on the right-hand side of the above equation contains the components of force along the coordinates r_1, r_2, r_3, etc. The gradient of the potential energy surface therefore determines the forces acting on all of the atoms, and thereby determines their subsequent motion. The potential energy surface (together with the initial conditions — whereabouts on the surface the system starts, and how much kinetic energy it has) therefore completely defines the reaction dynamics.

Conceptually, the collision trajectory can be thought of in terms of 'rolling a ball' over the potential energy surface, as shown schematically in Figure 5.11(b). As the ball rolls over the surface, different atomic configurations are sampled, allowing a 'movie' of the collision dynamics to be reconstructed. It is of course possible to be much more rigorous than this, and to calculate the trajectory of the system mathematically. To achieve this, starting from a particular point on the potential energy surface corresponding to the initial positions of the atoms, the gradient of the surface is calculated, which yields the forces acting on the atoms. Newton's second law of motion can therefore be solved in order to determine the trajectory, $\mathbf{r}(t)$, of the system, i.e. the way in which the atomic positions evolve as a function of time.

$$\mathbf{F} = m\mathbf{a} = m\frac{\mathrm{d}^2\mathbf{r}}{\mathrm{d}t^2} \qquad (5.36)$$

The system is allowed to follow this trajectory for a short period of time until it has reached a new point on the surface. It then becomes necessary to calculate the gradient again at this new position in order to determine an up-to-date set of forces, solve Equation (5.36) to find the new trajectory, let the system follow this trajectory for a short period of time to reach a new position on the surface, and so on. This procedure is known as a *classical trajectory calculation*, and with some modifications to take account of the quantised energy levels of the system (yielding so-called *quasi-classical trajectory* or QCT calculations), the approach may be used to predict virtually all of the experimental observables associated with a collision, including the identities of the products, their speed and angular distributions, and their quantum state populations. Such methods (and those described below) can also provide rate constants for comparison with experimental measurements or kinetic models.

If even more rigour is desired, fully quantum mechanical calculations can be performed in order to study the collision dynamics. This approach involves constructing an initial wavefunction for the system and propagating it across the potential energy surface by solving the time-dependent Schrödinger equation. As in the case of QCT calculations, described above, quantum scattering calculations can predict virtually any desired experimental observable for comparison with experimental data.

In the following, we will investigate the form of the potential energy surfaces for a variety of different chemical systems, and consider their consequences for the corresponding scattering processes.

5.4.15 The potential energy surface for a linear triatomic system

Consider a reacting system of three atoms, i.e. A + BC → AB + C. For simplicity, we will constrain the atoms to lie in a straight line throughout the reaction. This allows the relative positions of all three atoms to be defined by two distances, r_{AB} and r_{BC}, and the potential energy surface for the reaction is therefore a function of these two coordinates. To begin constructing the surface we first consider the interaction between the atoms in the *entrance channel* of the reaction, i.e. before the reactive collision, as the A atom approaches the BC molecule. Fixing the r_{BC} distance at some constant value corresponding to the equilibrium bond length in the reactant, there will be a long-range attractive interaction between A and BC, and a short-range repulsive interaction, and the potential energy curve will look something like that shown in Figure 5.12(a). The figure represents a cut through the full potential energy surface, $V(r_{AB}, r_{BC})$, for a fixed value of r_{BC}. The picture would be similar if we looked at the exit channel of the reaction, i.e. if we fixed the AB distance at the equilibrium bond distance for the AB product and plotted the interaction potential energy as a function of r_{BC}.

By considering the energy of the system at various A–B–C separations, the form of the complete PES can be predicted. Short-range repulsive interactions between A and BC result in the energy of interaction being high

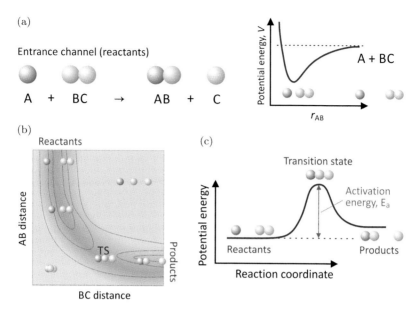

Fig. 5.12 (a) Potential energy curve plotted as a function of the r_{AB} distance for the entrance channel of the reaction A + BC; (b) schematic PES for a reaction A + BC → AB + C, shown as a contour plot. The atomic positions at various points on the surface are shown, and the transition state (TS) is marked; (c) The reaction coordinate and potential energy profile for a reaction A + BC → AB + C.

when either or both of the r_{AB} and r_{BC} distances are very small. It is also clear that the interaction energy will be lowest when r_{AB} and/or r_{BC} are at their equilibrium distance (corresponding to the minimum in the potential energy curve in Figure 5.12(a)). The energy increases again at large separations, where there are no attractive interactions between the atoms. The resulting potential energy surface is shown in Figure 5.12(b) as a contour plot. The surface resembles a curved 'half pipe'; keen skateboarders or snowboarders might be familiar with analogous gravitational potential energy surfaces of this form.

The minimum energy path across the surface follows the base of the 'half pipe', and is known as the *reaction coordinate*. Note that the transition state appears as a 'hump' along the reaction coordinate, which the reactants must surmount to form products. If the potential energy along the reaction coordinate is plotted, as shown in Figure 5.12(c), the familiar reaction potential energy profile for the reaction is recovered.

Having established the general form of the potential energy surface for a (linear) triatomic reaction, various different types of trajectory across the surface can now be explored.

5.4.16 Reactive and non-reactive trajectories across the potential energy surface

As noted earlier, the outcome of a collision depends both on the shape of the potential energy surface and on the initial velocities and internal states of the reactants. Investigating a few example trajectories across a surface similar to that constructed in Section 5.4.15, again for an A + BC → AB + C reaction, provides considerable insight into the connection between the topography of the potential energy surface and the dynamics of collisions governed by the surface. The four trajectories to be considered are shown in Figure 5.13.

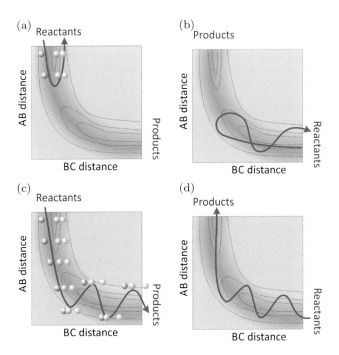

Fig. 5.13 Four different trajectories across the potential energy surface for an A + BC reaction. See text for explanation.

(A) The reactants are launched onto the surface with a low relative velocity. They have insufficient kinetic energy to surmount the activation barrier, and the trajectory 'turns round' and returns along the entrance channel to the reactants. Such a trajectory corresponds to elastic or inelastic scattering, depending on whether or not there is a change in rotational or vibrational state of the diatomic reactant.

(B) The reactants are launched onto the surface with a high relative velocity. They have sufficient energy to surmount the activation barrier, and at small AB distances they eventually roll up the repulsive wall of the surface, before rolling back down again and following an oscillating trajectory out into the product channel. Note that as atom C departs, the AB distance periodically extends and contracts, corresponding to vibration of the AB product.

From this simple consideration of two trajectories, we have therefore been able to predict that in order for reaction to occur on this potential energy surface, we need a significant amount of energy in relative translational motion of the reactants, and that some of this energy is converted into vibrational motion of the products.[15]

Trajectories C and D correspond to the reverse reaction, AB + C → A + BC.

(C) Trajectory C is another unsuccessful trajectory. The reactants have a reasonably high relative velocity in this case, but the position of the barrier in the exit channel for the reverse reaction means that the trajectory merely 'glances' off the barrier and returns along the entrance channel to reactants. In this case, we can see that some of the translational energy of the incoming reactants is converted into vibration of the AB molecule following the collision; this is therefore an inelastic scattering trajectory.

(D) In this trajectory the reactants begin with a mixture of translational and vibrational kinetic energy, and the trajectory successfully surmounts the barrier to form products with no vibrational excitation.

---

[15]This is a fairly common observation, first noted by John Polanyi, who received the Nobel Prize in Chemistry for his work on reaction dynamics in 1986. The observation forms the basis for 'Polanyi's rules', which link energy partitioning between vibration and translation in the products of a chemical reaction to the position of the barrier on the potential energy surface.

(*Note*: The description of trajectories above is somewhat simplified, and in reality the energy partitioning is often strongly dependent on the masses of the atoms involved in the reaction, as well as on the topography of the potential energy surface. The simple picture above can be recovered by plotting the potential energy surface in mass-weighted coordinates. This involves plotting the axes at a skew angle, β, given by

$$\cos^2 \beta = \frac{m_A m_C}{m_{AB} m_{BC}} \qquad (5.37)$$

for a reaction A + BC $\rightarrow$ AB + C, where m_X denotes the mass of species X. For interested readers, mass-weighted coordinates are treated in more detail in Appendix C).

5.4.17 General features of potential energy surfaces

The simple potential energy surface we have considered so far consist of reactant and product 'valleys' separated by a transition state. In general, potential energy surfaces can be much more complicated than this, particularly for larger chemical systems. In Figure 5.14, we define a number

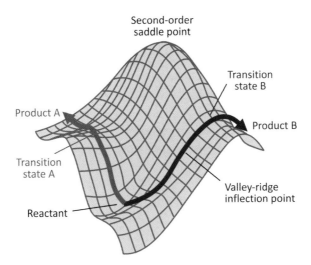

Fig. 5.14 Schematic of a general potential energy surface, with the various possible topological figures labelled. See text for explanation.

of different topological features that may be found on a reaction potential energy surface.[16] There are several features to note:

(1) Stable species appear as minima on the surface, with the deepest wells corresponding to the most stable species. Reactants and products tend to appear as fairly deep minima on the surface, while intermediates have shallower well depths.
(2) Transition states appear as saddle points on the surface. A saddle point (sometimes called a first-order saddle point) is a maximum in one dimension and a minimum in all other dimensions, and as the name suggests, is similar in shape to a horse's saddle.
(3) Other structures, such as second-order saddle points (a maximum in two dimensions and minimum in all others) and ridges, can also appear, but we will not look at these in any detail here.

5.4.18 Examples of potential energy surfaces for real chemical systems

We will now look at some examples of potential energy surfaces for a few bimolecular reactions and photoinitiated events. Note that not all of these processes are directly relevant to astrochemistry, but they all serve to illustrate the effects of various features of the potential energy surface on the dynamics of the process under study. Some of these processes involve more than one electronic state, and therefore more than one potential energy surface. As we will see, population can be transferred between different potential energy surfaces, opening up numerous possibilities for the reaction dynamics relative to processes occurring on a single surface.

5.4.18.1 *The simplest chemical reaction: $H + H_2 \to H_2 + H$*

The simplest chemical reaction, constrained to react in a linear geometry, has a perfectly symmetrical potential energy surface, with the transition state in the centre of the curve of the 'half pipe'. Figure 5.15(a) shows the surface in three-dimensional view and in contour view, with the reaction coordinate superimposed onto both views.

---

[16]The diagram is adapted from a figure appearing in a chapter on 'Geometry Optimisation' in the *Encyclopedia of Computational Chemistry* (John Wiley and Sons, 1998, pp. 1136–1142.

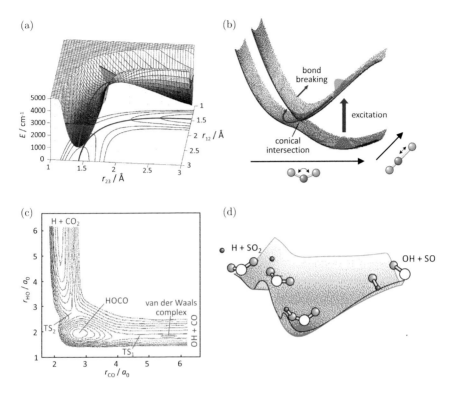

Fig. 5.15 (a) Potential energy surface for the reaction H + H$_2$ → H$_2$ + H, shown as both a contour plot and three-dimensional plot. The reaction coordinate is shown in blue, and dividing surfaces for reactants, transition state, and products are shown in purple. (b) Potential energy surface and dissociation mechanism for the photodissociation of NO$_2$. Potential energy surfaces for the reactions (c) H + CO$_2$ → OH + CO (contour plot) and (d) H + SO$_2$ (three-dimensional surface plot).

Source: (a) Adapted from *Acc. Chem. Res.*, **43**, 1519 (2010). (b) Adapted from *Science*, **331**, 411–412 (2011). (c) Adapted from *Science*, **334**, 208–212 (2011), and (d) *Chem. Phys. Lett.*, **433**(4–6), 279–286 (2007), respectively.

5.4.18.2 *Photodissociation of NO$_2$*

Figure 5.15(b) shows the ground and first electronically excited states for the NO$_2$ molecule, plotted as a function of the bond angle and one of the NO bond lengths (the second NO bond is held fixed). We can use the surfaces to understand the dissociation dynamics following absorption of a photon. Note that this process does not involve any collisions, though photodissociation is sometimes referred to as a 'half collision'. In NO$_2$, the equilibrium geometry for the excited state is much more bent than for the ground state, so following excitation the molecule bends to reduce its energy. The excited and ground-states cross, forming a feature known as a conical intersection. The wavefunction can funnel through the conical intersection back to highly excited vibrational levels of the ground state, from which the molecule dissociates. The bent geometry from which dissociation occurs leads to a high degree of rotational excitation in the diatomic fragment, since the N atom receives a strong 'kick' when the N–O bond dissociates.

5.4.18.3 $H + CO_2 \rightarrow OH + CO$ and $OH + SO \rightarrow H + SO_2$

In contrast to the systems we have considered so far, the $H + CO_2$ reaction has a deep well on its potential energy surface, corresponding to the relatively stable molecule HOCO (see Figure 5.15(c)). This leads to very different collision dynamics from those of the examples presented above. Unlike the direct dynamics we have considered so far, in which molecules collide, exchange atoms, and separate in essentially a single concerted step, when a deep well is present on the surface the system can become trapped in the well for significant periods of time. Direct reactions often lead to strongly anisotropic product angular distributions, with a strong preference for forward or backward scattering depending on whether the potential energy surface steers the reaction towards 'glancing' or 'head-on' collisions, respectively. In contrast, when a long-lived complex forms and survives for multiple rotational periods, all 'memory' of the initial approach directions of the reactants is lost, and an isotropic angular distribution results. The energy available to the system also has time to become distributed statistically amongst the various modes of motion of the complex, leading to characteristic 'statistical' population distributions in the reaction products. In contrast, the newly-formed (or 'nascent') products of direct reactions often have highly non-statistical quantum state distributions.

As we might expect, the situation is similar for the $H + SO_2$ reaction, in which the carbon atom has been replaced by sulphur. A three-dimensional view of the potential energy surface is shown in Figure 5.15(d).

5.4.18.4 The $Ar + H_2^+ \rightarrow ArH^+ + H$ reaction

Finally, we will take a brief look at a potential energy surface for an ion–molecule reaction. Figure 5.16 shows cuts through the potential energy surface for the ArH_2^+ system at various different Ar–H–H angles. As can be seen for the plots at angles of 180° and 135°, there is no barrier to reaction at linear and slightly bent geometries, and there is a shallow well on the surface corresponding to the $Ar \cdot H_2^+$ complex. At smaller angles, there are two minima on the surface, separated by a small barrier.

QCT calculations have been carried out over the surface[17] in order to determine the reaction cross-section, and the results reproduce the features we would expect for an ion–molecule reaction (see Section 5.4.8). At low collision energies (1 eV), where there is plenty of time for the intermolecular interactions to influence the collision dynamics, the cross-section is relatively large, at between 15 and 30 Å^2, depending on H_2 vibrational state. There is relatively little dependence on vibrational state, as expected for a

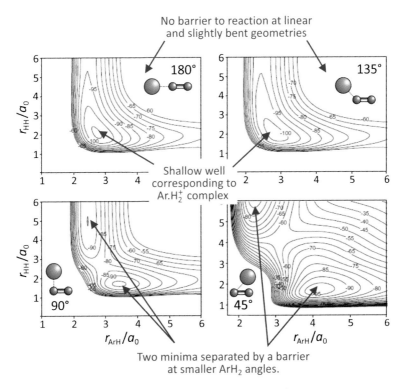

Fig. 5.16 Potential energy surface for the reaction $Ar + H_2^+ \rightarrow ArH^+ + H$ at various different Ar–H–H angles.

Source: Adapted from Liu *et al.* (2011).

---

[17] X. Liu, H. Liu, and Q. Zhang, *Chem. Phys. Lett.*, **507**, 24–28 (2011).

barrierless reaction. At higher collision energies, the cross-section reduces sharply, reaching a value of around 4 Å^2 (independent of H$_2$ vibrational state) at a collision energy of 6 eV.

5.4.19 Orbital angular momentum, centrifugal barriers, and the effective potential

In Section 5.4.9, we defined the orbital angular momentum of a pair of colliding particles, and noted that orbital angular momentum is conserved during a collision. Consider again the particle trajectory shown in Figure 5.8. The relative kinetic energy of the two particles, $\frac{1}{2}\mu v_{\text{rel}}^2$, where μ is the reduced mass of the two particles, and v_{rel} is their relative velocity, can be expressed as a sum of translational kinetic energy directed along the line of centres **r** of the two particles and rotational kinetic energy associated with the orbital motion described above, i.e.

$$K_{\text{rel}} = \frac{1}{2}\mu v_{\text{rel}}^2 = \frac{1}{2}\mu \dot{r}^2 + \frac{L^2}{2\mu r^2} \tag{5.38}$$

where $\dot{r}$ denotes the rate of change of the 'line-of-centres distance' or particle separation with time. Only the component of the kinetic energy associated with motion along the line of centres is available to help overcome an activation barrier to reaction. Because the total angular momentum, **L**, must be conserved during the collision, the kinetic energy associated with the orbital motion is not available to help surmount an activation barrier. The orbital kinetic energy therefore has the effect of reducing the available energy, and as a result, this term is often referred to as a *centrifugal barrier*. The centrifugal barrier term is often combined with the potential energy surface to give an effective potential $V_{\text{eff}}(r)$, i.e.

$$V_{\text{eff}}(r) = V(r) + \frac{L}{2\mu r^2} \tag{5.39}$$

As shown in Figure 5.17, the centrifugal barrier can give rise to an effective barrier to reaction, even when the potential energy surface itself has no barrier. Using $L^2 = (\mu v_{\text{rel}} b)^2$ (see Section 5.4.9), we can rewrite the effective potential as

$$V_{\text{eff}}(r) = V(r) + \frac{1}{2} \frac{\mu v_{\text{rel}}^2 b^2}{r^2} \tag{5.40}$$

We see that the barrier is largest for heavy systems, particularly when colliding at large impact parameters. The centrifugal barrier often has the

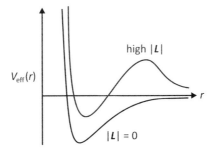

Fig. 5.17 The effective potential for low and high values of the impact parameter, b, and therefore the orbital angular momentum, L, showing the appearance of a centrifugal barrier at large values of the orbital angular momentum.

effect of reducing the maximum impact parameter that can lead to successful reaction, thereby reducing the reaction cross-section.

5.4.20 A simple model for the rate of ion–molecule reactions

We have already alluded to the fact that state-of-the-art QCT and quantum scattering calculations are capable of predicting reaction cross-sections and rate constants for ion–molecule reactions. However, these calculations are only possible when a detailed potential energy surface is available for the reaction under study, which is often not the case. In the following, we will show how we can use the basic principles of collision physics covered in Sections 5.4.2 onwards to understand the key factors determining the rates of ion–molecule reactions, and to develop a simple expression which can be used to calculate approximate rate constants. The model we will derive is known as the *Langevin model*[18] (pronounced 'Lon-je-van'), and is the simplest of a variety of capture theories that have been developed to describe the rates of ion–molecule reactions.

We start from the assumption that reaction is governed by the long-range part of the interaction potential, a region that can be treated classically to a reasonable approximation. The effective potential in this region will contain contributions from ion–induced-dipole interactions, ion–permanent dipole interactions (if the neutral molecule has a permanent

---

[18]P. Langevin, *Ann. Chim. Phys.*, **5**, 245 (1905).

dipole), ion-quadrupole interactions, and so on, as well as a term representing the centrifugal barrier. We will consider only the simplest case, in which the molecule has no permanent dipole, and will include only the contributions from the ion–induced-dipole interaction (see Table 5.1) and the centrifugal barrier to the effective potential. The long-range potential as a function of the ion–molecule separation r is then

$$V_{\text{eff}}(r) = -\frac{\alpha q^2}{8\pi\epsilon_0 r^4} + \frac{\mu v_{\text{rel}}^2 b^2}{2r^2} \qquad (5.41)$$

In this expression, α is the polarisability of the neutral molecule, q is the charge on the ion, and ϵ_0 is the permittivity of free space.

As the reactants approach, they initially have a kinetic energy of relative motion equal to $K_{\text{rel}} = \frac{1}{2}\mu v_{\text{rel}}^2$. As they move closer and begin to experience the centrifugal barrier, some of this kinetic energy is converted into potential energy. In order for the reactants to surmount the centrifugal barrier and for reaction to take place, the initial kinetic energy must be greater than the barrier height, i.e. $K_{\text{rel}} > V_{\text{eff}}(r_{\text{max}})$, where r_{max} is the position of the centrifugal barrier. This condition is shown schematically in Figure 5.18(a) for a number of different impact parameters, b. Because the height of the centrifugal barrier is determined by the impact parameter, there will be a maximum impact parameter b_{max} beyond which the particles do not have sufficient energy to react. We shall show in the following that by determining b_{max}, we can calculate a value for the rate constant.

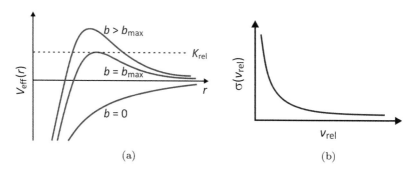

Fig. 5.18 (a) Since the height of the centrifugal barrier depends on the initial impact parameter, b, there will be a maximum impact parameter b_{max} beyond which the reactants do not have sufficient kinetic energy to overcome the barrier and react; (b) The Langevin reaction cross-section as a function of reactant relative velocity v_{rel}.

To find the value of r for which $V_{\text{eff}}(r)$ is a maximum, we solve

$$\frac{dV_{\text{eff}}(r)}{dr} = 0 \tag{5.42}$$

After some algebra (left as a straightforward exercise for the reader), we obtain

$$r_{\max} = \left(\frac{\alpha q^2}{2\pi\epsilon_0 \mu v_{\text{rel}}^2 b^2}\right)^{1/2} \tag{5.43}$$

The barrier height is therefore

$$V_{\text{eff}}(r_{\max}) = -\frac{\alpha q^2}{8\pi\epsilon_0 r_{\max}^4} + \frac{\mu v_{\text{rel}}^2 b^2}{2 r_{\max}^2}$$

$$= \frac{\pi\epsilon_0 \mu^2 v_{\text{rel}}^4 b^4}{2\alpha q^2} \tag{5.44}$$

For reaction to occur, we require $K_{\text{rel}} \geq V_{\text{eff}}(r_{\max})$

$$\frac{1}{2}\mu v_{\text{rel}}^2 \geq \frac{\pi\epsilon_0 \mu^2 v_{\text{rel}}^4 b^4}{2\alpha q^2} \tag{5.45}$$

Rearranging, we find that the maximum impact parameter for which reaction can occur is given by

$$b^2 \leq \left(\frac{\alpha q^2}{\pi\epsilon_0 \mu v_{\text{rel}}^2}\right)^{1/2} \tag{5.46}$$

The reaction cross-section is then

$$\sigma_r(v_{\text{rel}}) = \pi b_{\max}^2 = \pi \left(\frac{\alpha q^2}{\pi\epsilon_0 \mu v_{\text{rel}}^2}\right)^{1/2} \tag{5.47}$$

According to this expression (sometimes referred to as the *Langevin cross-section*), the reaction cross-section has a $1/v_{\text{rel}}$ dependence on the reactant relative velocity, and takes the form shown in Figure 5.18(b). Note that although the reactant kinetic energy increases with v_{rel}, so too does the centrifugal barrier, with the net effect being a dramatic decrease in cross-section as the relative velocity increases. We see that the velocity-dependence of our reaction cross-section reproduces that predicted by our qualitative description of the excitation function for a barrierless reaction in Section 5.4.8.

As demonstrated in Section 5.4.6, the rate constant for a given relative velocity is given by $k(v_{\text{rel}}) = \sigma_{\text{r}}(v_{\text{rel}})v_{\text{rel}}$. The thermal rate constant can therefore be found by integrating the rate constant over the Maxwell–Boltzmann distribution of relative velocities, $f(v_{\text{rel}})$ (assuming a thermal distribution of velocities), to give

$$k(T) = \int_0^\infty v_{\text{rel}} \sigma_{\text{r}}(v_{\text{rel}}) f(v_{\text{rel}}) \mathrm{d}v_{\text{rel}}$$

$$= \int_0^\infty v_{\text{rel}} \pi \left(\frac{\alpha q^2}{\pi \epsilon_0 \mu v_{\text{rel}}^2}\right)^{1/2} f(v_{\text{rel}}) \mathrm{d}v_{\text{rel}} \quad \text{(substituting for } \sigma_{\text{r}}(v_{\text{vel}})\text{)}$$

$$= \left(\frac{\pi \alpha q^2}{\epsilon_0 \mu}\right)^{1/2} \int_0^\infty f(v_{\text{rel}}) \mathrm{d}v_{\text{rel}}$$

$$= \left(\frac{\pi \alpha q^2}{\epsilon_0 \mu}\right)^{1/2} \quad \text{(since } f(v_{\text{rel}}) \text{ is normalised)} \tag{5.48}$$

Note that because we have only considered the possibility of a centrifugal barrier to reaction (i.e. we have ignored any true barriers that may be present on the potential energy surface), this expression represents an upper limit to the rate coefficient for an ion–molecule reaction. A key point to note is that the thermal rate constant we have derived is *independent of temperature*, in line with our previous discussion of ion–molecule reactions.

There are many more sophisticated models available for modelling ion–molecule reaction rates. These include quantum mechanical models such as the impressively-named 'adiabatic capture and centrifugal sudden approximation' (ACCSA) theory,[19] which takes into account the rotational states of the reactants; variational transition state theory[20]; and trajectory calculations. The latter two approaches require a reasonably detailed knowledge of the potential energy surface for the reaction. Variational transition state theory is an improved version of transition state theory, which you may have come across in statistical mechanics courses. The basics of trajectory

---

[19] D. C. Clary, Calculations of rate constants for ion–molecule reactions using a combined capture and centrifugal sudden approximation, *Mol. Phys.*, **54**(3), 605–618 (1984).
[20] D. Truhlar and B. Garret, Variational transition state theory, *Annu. Rev. Phys. Chem.*, **35**, 159–189 (1984).

calculations were covered in Section 5.4.14. To round off this chapter, we will look briefly at employing trajectory calculations to determine reaction cross-sections.

5.4.21 Reaction cross-sections from quasi-classical trajectory calculations

Determining reaction cross-sections, and therefore thermal rate constants, from classical trajectory or QCT data is in principle relatively straightforward. The reaction probability $P_r(E_{coll}, v, J, b)$ for a specified collision energy, E_{coll}, reactant rotational and vibrational state (v, J), and impact parameter, b, is simply the ratio of the number of reactive trajectories, N_r, to the total number of trajectories, N.

$$P_r(E_{coll}, v, J, b) = \lim_{N \to \infty} \frac{N_r(E_{coll}, v, J, b)}{N(E_{coll}, v, J, b)} \tag{5.49}$$

When running the trajectories, it will be found that there is a maximum impact parameter, b_{max}, above which reaction does not occur. The reaction cross-section is then given by (see Section 5.4.7)

$$\sigma_r(E_{coll}, v, J) = \int_0^{b_{max}} P_r(E_{coll}, v, J, b) 2\pi b \, db \tag{5.50}$$

To determine the thermal rate constant, we sum over the initial quantum states v and J to determine the total reaction cross-section as a function of collision energy, i.e. the excitation function. We can then integrate over the distribution of collision energies to determine the thermal rate constant, as described in Section 5.4.8.

5.5 Summary

We have explored in some detail the way in which theoretical methods can be used to predict molecular structures, spectra, and reactivities for comparison with experimental and observational data on interstellar molecules. In Chapter 6, we will outline a variety of experimental approaches that can be used to measure spectra for the types of unstable molecules found in interstellar space, together with a further set of techniques that can be used to characterise their reactivity.

5.6 Questions

5.6.1 Essay-style questions

Q5.1 Explain why homonuclear diatomics cannot be detected via microwave or infrared spectroscopy.

Q5.2 What types of information can be obtained from the microwave spectrum of a molecule in an interstellar gas cloud?

Q5.3 Using theoretical methods only, how might one go about identifying the chemical species giving rise to a set of unassigned lines in a microwave spectrum recorded from an interstellar gas cloud?

Q5.4 Explain why reaction cross-sections tend to decrease with increasing collision energy.

Q5.5 Outline the basic principles of angular momentum conservation in reactive collisions.

Q5.6 Explain how the PES for a reaction determines the outcome of a collision.

5.6.2 Problems

P5.1 *Simple variational electronic structure calculation on H_2*

You may find it helpful to refer to Appendix B when answering this problem.

A simple trial wavefunction for the nth molecular orbital of the H_2 molecule takes the form

$$\psi_n = c_{1n}\phi_1 + c_{2n}\phi_2$$

where ϕ_1 and ϕ_2 are 1s orbitals on the two H atoms, and c_{1n} and c_{2n} are constants. The energy of the orbital is given by

$$E_n = \frac{\langle \Psi|\hat{H}|\Psi_n\rangle}{\langle \Psi_n|\psi_n\rangle}$$

(a) By substituting the expression for ψ_n into the expression for the energy, show that

$$E_n(c_{1n}^2 S_{11} + c_{2n}^2 S_{22} + 2c_{1n}c_{2n}S_{12})$$
$$= c_{1n}^2 H_{11} + c_{2n}^2 H_{22} + 2c_{1n}c_{2n}H_{12}$$

where

$$H_{ij} = \langle \phi_i|\hat{H}|\phi_j\rangle, \quad H_{ij} = H_{ji}$$
$$S_{ji} = \langle \phi_i|\phi_j\rangle, \quad S_{ij} = S_{ij}$$

(b) Show that by applying the variation principle, i.e. $dE_n/dc_i = 0$, one obtains a pair of secular equations.

$$c_{1n}(H_{11} - E_n S_{11}) + c_{2n}(H_{12} - E_n S_{12}) = 0$$
$$c_{1n}(H_{21} - E S_{21}) + c_{2n}(H_{22} - E_n S_{22}) = 0$$

(c) For H_2, we define the quantities $\alpha = H_{11} = H_{22}$, and $\beta = H_{12} = H_{21}$ (note that both α and β are negative quantities) and use the simplification that $S_{ij} = \delta_{ij}$ (i.e. $S_{11} = S_{22} = 1$, $S_{12} = S_{21} = 0$). Show that with these simplifications, the secular equations can be written in matrix form as follows:

$$\begin{pmatrix} \alpha - E_n & \beta \\ \beta & \alpha - E_n \end{pmatrix} \begin{pmatrix} c_{1n} \\ c_{2n} \end{pmatrix} = 0$$

(d) The allowed energy levels are found by setting the determinant of the matrix equal to zero and solving for E. Show that this leads to energies

$$E_1 = \alpha + \beta \quad E_1 = \alpha - \beta$$

(e) The wavefunction corresponding to each energy is found by substituting each energy into the secular equations in turn, and solving for the coefficients. This allows all of the coefficients to be determined in terms of one coefficient, and normalisation of the wavefunction allows the final coefficient to be determined. Show that the optimum trial wavefunctions for H_2 are

$$\Psi_1 = \frac{1}{\sqrt{2}}(\phi_1 + \phi_2) \quad \text{and} \quad \Psi_1 = \frac{1}{\sqrt{2}}(\phi_1 - \phi_2)$$

(f) Comment on this result in terms of your knowledge of the bonding in H_2.

P5.2 *Gross selection rules for microwave and infrared spectroscopy*
Which of the following molecules have: (a) a microwave spectrum; (b) an infrared spectrum?

$$H_2, C_2, CN, CO, HCl, HCN, H_2S, SO_2,$$
$$CO_2, NH_3, CH_3, CH_4, CH_3CN, C_2H_4$$

P5.3 *Moment of inertia of a diatomic molecule*
Prove that for a diatomic molecule, Equations (5.3) and (5.4) are equivalent.

P5.4 *Rotational populations of CO and in an interstellar gas cloud*
The rotational constant of CO is 1.9313 cm^{-1}. Determine the most populated rotational level of CO in an interstellar gas cloud at temperatures of (a) 10 K and (b) 100 K.

P5.5 *Force constants and vibrational frequencies*
H$_2$, D$_2$, O$_2$, and N$_2$ have vibrational wavenumbers of 4401.21, 3115.50, 1580.19, and 2358.57 cm^{-1}, respectively. Assuming harmonic vibrational motion, calculate the force constant for each molecule, and comment on your results in light of the bonding in each of the four molecules.

P5.6 *Vibrational degrees of freedom in polyatomics*
An N-atom molecule has $3N$ degrees of freedom in total. By considering the number of degrees of freedom in translational and rotational motions, determine the number of vibrational modes in the following molecules:

$$H_2, CO_2, H_2O, NH_3, CH_4, C_2H_2, C_{60}$$

P5.7 *Kinetic energy in the lab and centre-of-mass frames*
Prove that Equations (5.19) and (5.20) are equivalent.

P5.8 *Dynamics of the ion–molecule reaction $C^+ + OH \to CO^+ + H$*

The following data relate to the reaction $C^+ + OH \to CO^+ + H$

Enthalpies of formation at 0 K: $\Delta_f H(C^+) = 1802.99 \text{ kJ mol}^{-1}$

$\Delta_f H(OH) = 38.99 \text{ kJ mol}^{-1}$

$\Delta_f H(CO^+) = 1241.43 \text{ kJ mol}^{-1}$

$\Delta_f H(H) = 218.00 \text{ kJ mol}^{-1}$

Polarisability volume:	$\alpha(OH) = 1.07 \text{ Å}^3$
Vibrational wavenumber:	$\omega(CO^+) = 2214.2 \text{ cm}^{-1}$
Rotational constant:	$B(CO^+) = 1.9772 \text{ cm}^{-1}$

(a) Calculate the mean relative velocity of C^+ and OH in an interstellar gas cloud at a temperature of 10 K, assuming a Maxwell–Boltzmann distribution of velocities.
(*Note*: You may use $v_{\text{rel}} = (8k_BT/\pi\mu)^{1/2}$, where T is the temperature, and μ the reduced mass of the collision partners).

(b) Determine the maximum impact parameter for the collision at the mean relative velocity calculated in (a), and hence the Langevin reaction cross-section.

(c) Determine the orbital angular momentum of the reactants if they approach at the maximum allowable impact parameter.

(d) Calculate the reaction enthalpy, and the energy available to the reaction products, assuming that the reactants possess no internal energy.

(e) Assuming all of the initial orbital angular momentum is transformed into rotational angular momentum of the CO^+ product (a good approximation for this system), determine the most probable rotational state of this product. You may ignore centrifugal distortion.

(f) How much energy remains for vibrational and translational excitation of the products? What is the highest vibrational level of CO^+ that may be populated in the reaction? You may ignore anharmonicity.

P5.9 *Molecular interactions in an ion–molecule collision*

H_2O^+ abstracts a proton from H_2O in the reaction $H_2O^+ + H_2O \rightarrow H_3O^+ + OH$. Assuming a temperature of 50 K, use the data below (dipole moment μ, polarizability volume α, and ionisation potential I), together with Equation (5.33) and Table 5.1, to determine the ion–dipole, dipole–dipole, ion–induced dipole, dipole–induced dipole, and induced dipole–induced dipole contributions to the attractive potential between the two reactants at a separation of 3 Å. Comment on your answer.

H_2O^+ $\quad \mu = 2.3602 D \quad$ H_2O $\quad \mu = 1.8550 D$
$\quad \alpha = 0.628 \text{Å}^3 \quad\quad\quad\quad \alpha = 1.470 \text{Å}^3$
$\quad I = 38.63 \text{ eV} \quad\quad\quad\quad\; I = 12.621 \text{ eV}$

(*Note*: 1 Debye (D) $= 3.33564 \times 10^{-30}$ C m.
1Å $= 1 \times 10^{-10}$ m).

P5.10 *Derivation of the Langevin model for ion–molecule reaction rates*

A simple classical one-dimensional model for the collision-energy dependence of the reaction cross-section uses the following expression:

$$E_{\text{coll}} = \frac{1}{2}\mu \dot{R}^2 + V_{\text{eff}}(R)$$

where E_{coll} is the initial collision energy and $V_{eff}(R)$ is the radial dependence of the effective potential.

(a) Explain the origins of this expression, and present arguments to show that the effective potential may be written as

$$V_{eff}(R) = V(R) + E_{coll}\frac{b^2}{R^2} \tag{1}$$

where $V(R)$ is the radial dependence of the potential energy, and b is the impact parameter for the collision. Recall that the kinetic energy associated with orbital motion can be written as

$$E_{cent} = \frac{L^2}{2\mu R^2}$$

(b) The long-range interaction potential for a reaction without a barrier has the general form

$$V(R) = -\frac{C_n}{R^n}$$

where $n > 1$, and C_n is a constant that depends on the electrostatic properties (charges, dipole moments, etc.) of the reactants. Use this expression, together with Equation (1), to show that the location of the maximum in the centrifugal barrier on the effective potential is given by the expression

$$R_0 = \left(\frac{nC_n}{2b^2 E_{coll}}\right)^{\frac{1}{n-2}}$$

(c) For a reaction between an ion and a molecule, the ion–induced dipole interaction has $n = 4$. Show that for the radial kinetic energy at the barrier to be greater than zero, the impact parameter for this type of reaction obeys the following inequality.

$$b^2 \leq \left(\frac{4C_4}{E_{coll}}\right)^{\frac{1}{2}}$$

(d) Write down an expression for the reaction cross-section, $\sigma(v_{rel})$, stating any additional assumptions made, and sketch the dependence of the cross-section on collision energy (i.e. sketch the excitation function).

(e) The thermal rate constant can be written as

$$k(T) = \int_0^\infty v_{\text{rel}} \sigma(v_{\text{rel}}) f(v_{\text{rel}}) dv_{\text{rel}}$$

where $f(v_{\text{rel}})$ is the Maxwell–Boltzmann distribution of relative velocities, v_{rel}, which may be assumed to be normalised, and $\sigma(v_{\text{rel}})$ is the cross-section determined in part (d), but expressed in terms of relative velocity. Determine an expression for the thermal rate coefficient for an ion–molecule reaction, and comment on the result you obtain.

Chapter 6

Laboratory-Based Astrochemistry: Experiment

6.1 Experimental astrochemistry I: Spectroscopic data

As noted previously, identifying molecules in the interstellar medium and other regions of space requires knowledge of their spectra. In Section 5.3, we learnt how to calculate rotational, vibrational, and electronic spectra from first principles. Calculations of this type are relatively straightforward to perform for small molecules, but scale in size dramatically as the number of electrons in the molecule increase. Calculating accurate spectra for large interstellar molecules from first principles remains extremely challenging.

If it is possible to generate the molecule of interest in sufficient quantities in the laboratory, a spectroscopic study provides experimental data that can be compared directly with observational data from telescopes, once the appropriate corrections have been made for Döppler shift. The spectral region of most interest for molecular identification is the microwave region, which provides information on the rotational energy levels of the molecule. Conveniently, this region coincides with an 'atmospheric window', allowing molecular species to be observed using ground-based telescopes. At room temperature, a large number of rotational energy levels are populated, and rotational spectra are often extremely complex. In contrast, in the interstellar medium, the low ambient temperatures result in only a few rotational quantum states being occupied. The dramatic effect of such a reduction in temperature on the rotational spectrum of a medium-sized organic molecule is shown in Figure 6.1. Ideally, we would like to cool our molecules down to

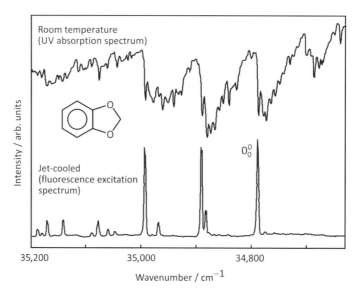

Fig. 6.1 Reducing the temperature of a sample reduces the number of occupied rotational levels, often yielding a dramatic simplification in the rotational spectrum.

Source: Adapted from Laane *et al.*, *J. Mol. Struct.*, **480–481**, 189–196 (1999).

a temperature of a few Kelvin, in order to record spectra under conditions that mimic the interstellar medium as closely as possible. This is usually achieved by preparing the sample in a molecular beam. In the following, we will describe how molecular beams are generated and summarise their properties, before looking at a number of spectroscopic techniques that can be used to record molecular spectra.

6.1.1 Molecular beams

There are two general categories of molecular beam source, known as *effusive* and *supersonic* sources, respectively. Both types of source work by allowing gas to escape from a high pressure region through a small orifice into a vacuum. The flow characteristics of the source are determined by the ratio of the mean free path λ of the molecules in the source (defined as the mean distance the molecules travel between collisions) to the diameter ϕ of the orifice, a quantity known as the Knudsen number, K_n

$$K_n = \frac{\lambda}{\phi} \tag{6.1}$$

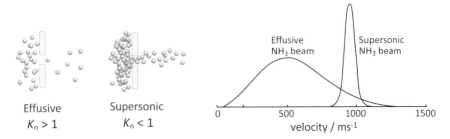

Fig. 6.2 Preparation of effusive and supersonic molecular beams, and the resulting molecular speed distributions. See text for details.

In an effusive source, the diameter of the orifice is smaller than the mean free path in the gas ($K_n > 1$), and in a supersonic source it is larger ($K_n < 1$). The orifice size is generally similar in the two types of source, but the supersonic source operates at a much higher gas pressure, resulting in a much shorter mean free path than in the effusive source. This is shown schematically in Figure 6.2, together with the beam velocity distributions for effusive and supersonic beams of a representative interstellar (and terrestrial) molecule, NH_3. It is immediately apparent from these distributions that the source pressure has an enormous effect on the beam properties. Effusive sources give rise to a broad velocity distribution peaking at relatively low velocity, while supersonic sources give rise to a narrow distribution peaking at much higher velocity. These characteristic velocity distributions arise as a result of the markedly different collisional regimes within the orifice and immediately beyond it in the two cases, as explained in the following.

6.1.1.1 *Effusive sources*

In an effusive beam, molecules effectively 'wander' out of the orifice whenever they 'collide' with it, with the result that the Maxwell–Boltzmann distribution of molecular speeds in the source is more or less maintained in the molecular beam. In fact, the distribution is somewhat skewed towards higher velocities, since molecules with higher speeds undergo more frequent collisions with the walls and are therefore more likely to exit the orifice. The beam has a broad $\cos^2 \theta$ angular distribution about the beam axis (i.e. the direction normal to the wall of the chamber containing the orifice), since the random velocity components of the molecules within the source are conserved as they exit through the orifice. As noted above, effusive sources contain the gas at a low pressure, and are generally only used to produce beams of metal atoms or other species that can only be prepared at low pressure in the gas phase. Usually, the source is heated to high temperatures in order to obtain a vapour pressure as high as possible. Such sources are clearly not very useful for low-temperature spectroscopy studies.

6.1.1.2 *Supersonic sources*

In a supersonic source, because the mean free path is much smaller than the diameter of the orifice, many collisions occur as the molecules exit the orifice and in the region immediately beyond it. Collisions that impart a velocity component along the beam axis are most successful at allowing a molecule to escape this region, with the result that the molecules entering the beam are those for which the collisions have converted almost all of their random translational energy and internal (rotational and vibrational) energy into directed translational kinetic energy along the beam axis. The beam molecules therefore have almost no internal energy, occupying only very low rotational quantum states, and have a very narrow speed distribution. The angular distribution about the beam axis is also much narrower than for an effusive beam. Since the width of the molecular speed distribution determines the temperature of a gas, by this definition the molecules in a supersonic molecular beam are extremely cold. It is common to reach temperatures as low as 5 K by this very simple technique of expanding a gas through a small hole. At these low temperatures, only a few low-energy rotational states are occupied. Vibrational excitation is less efficiently cooled during the expansion, and the possible presence of vibrationally excited molecules within the beam should be borne in mind when analysing and interpreting data from molecular beam experiments.

Unstable species can be created at low temperatures within a supersonic expansion by dissociating a suitable precursor molecule. This is usually achieved either by applying a high-voltage discharge as the beam exits the source, or by crossing the beam with an ultraviolet (UV) laser beam of an appropriate wavelength immediately after the source. The newly-formed radical species of interest are then cooled during the large number of collisions that occur within a few nozzle diameters of the source, and become entrained within the molecular beam.

The maximum or terminal velocity of molecules within a supersonic beam can be determined by assuming that all of the internal energy of the molecules inside the source is converted into kinetic energy directed along the beam axis, in an *isenthalpic* expansion (i.e. $\Delta H = 0$ during the expansion). Assuming ideal gas behaviour, we have

$$\frac{1}{2}mu^2 = -\int_{T_s}^{T_f} C_p \mathrm{d}T \tag{6.2}$$

where $\frac{1}{2}mu^2$ is the kinetic energy of the beam molecules with mass m and speed u, and $C_p \mathrm{d}T = \mathrm{d}H$ is the change in enthalpy of the molecules within the source as they undergo supersonic expansion (C_p is the heat capacity of the molecules at constant pressure). T_s and T_f are the temperatures of the source and molecular beam, respectively. Since T_f is small relative to T_s, we have $T_s - T_f \approx T_s$. Also, assuming C_p is not temperature dependent, we can set $C_p = \gamma R/(\gamma - 1)$, where $\gamma = C_p/C_V$. Evaluating the right-hand side of Equation (6.2) then gives

$$\frac{1}{2}mu^2 = \left(\frac{\gamma R}{\gamma - 1}\right) T_s \tag{6.3}$$

Rearranging yields the terminal velocity of the beam molecules.

$$u = \left(\frac{2RT_s}{m} \frac{\gamma}{(\gamma - 1)}\right)^{1/2} \tag{6.4}$$

We see from Equation (6.4) that the terminal velocity of a supersonic molecular beam depends on the mass of the beam gas, with light gases achieving higher speeds than heavier gases. For example, a beam of He has a terminal velocity of around $2400\,\mathrm{m\,s^{-1}}$, while for HCl it is much lower, at around $500\,\mathrm{m\,s^{-1}}$. Higher beam velocities for heavier molecules may be achieved by 'seeding' the heavier molecule in a light carrier gas so as to achieve a lighter 'average mass' for the beam molecules; a beam of 1% HCl in He, for example, has a terminal velocity of around $2100\,\mathrm{m\,s^{-1}}$.

By definition, supersonic beams travel faster than the local speed of sound, resulting in a shock-wave structure (see Figure 6.3(a)) consisting primarily of a 'barrel shock' wave that surrounds the beam and a 'Mach disk' that intersects the beam axis. Immediately in front of the nozzle orifice is a 'zone of silence', which is not disturbed by shock waves. By placing a cone-shaped skimmer (see Figure 6.3(b)) in the zone of silence, 1–2 cm from the nozzle, it is possible to create a collimated beam which is not disturbed by turbulence from the shock waves. As shown in Figure 6.3(c), the skimmer is shaped so as to pass molecules from the core of the beam,

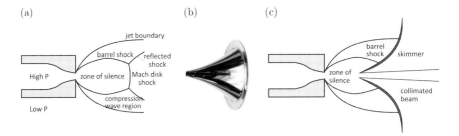

Fig. 6.3 (a) Shock-wave structure of a supersonic molecular beam; (b) an electroformed nickel skimmer; (c) use of a skimmer to deflect supersonic shock waves and to collimate the molecular beam.

while deflecting the barrel shock wave and preventing the Mach disk from forming.

6.1.2 Fourier-transform microwave spectroscopy

Fourier-transform (FT) microwave spectroscopy is one of the most widely used methods for astrochemical spectroscopy studies, offering extremely high resolution at relatively low cost. A schematic of a Fourier-transform spectrometer[1] is shown in Figure 6.4.

The molecular beam containing the species of interest enters a Fabry–Perot (two-mirror) cavity. It is then irradiated by a microwave pulse, which forms a standing wave within the cavity. The pulse excites molecules within the beam to higher rotational states, and also creates spatial coherence between their rotational motion, so that there is a well-defined phase relationship between the rotational motion of molecules in different rotational states. After the pulse, the molecules relax back to the ground state and emit microwave radiation coherently at their resonant frequencies as they do so, producing an interference pattern. The interference pattern is detected using a superheterodyne detector,[2] and is then Fourier transformed to obtain the rotational spectrum.

---

[1] The first microwave Fourier-transform spectrometer was built by Jan Ekkers and W. H. Flygare in 1975, and is described in their publication 'Pulsed microwave Fourier transform spectrometer', *Rev. Sci. Instrum.*, **47**, 448 (1976).

[2] Heterodyne detection shifts the detected signal to higher frequencies by mixing it with a carrier frequency, thereby improving the signal-to-noise ratio of the measurement.

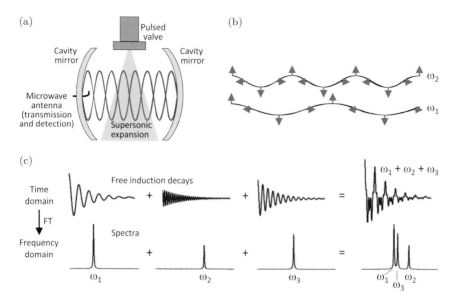

Fig. 6.4 (a) Schematic of a Fourier-transform microwave spectrometer — see text for details; (b) the microwave excitation pulse excites the beam molecules coherently into a variety of rotational states with a well-defined phase relationship between their rotational motion; (c) the excited molecules emit radiation coherently to return to the ground state, resulting in an interference pattern or free induction decay in the time domain, which can be Fourier transformed into the frequency domain to yield the rotational spectrum.

6.1.3 Laser-induced fluorescence

Laser-induced fluorescence, often abbreviated as 'LIF', is a laser spectroscopy technique that can be used to record spectra of molecules possessing a strongly fluorescing excited electronic state. The fluorescent state must be optically accessible from the ground state. The technique is extremely sensitive, and is therefore well suited to measurements on the relatively low densities of molecules within molecular beams. The technique is illustrated schematically in Figure 6.5.

As previously, the molecule to be studied is prepared in a supersonic expansion. The molecular beam is then crossed by a laser beam, which excites molecules from the ground to the fluorescent excited electronic state. The photons emitted from the excited state are then detected by a photomultiplier tube or other sensitive photodetector. By scanning the excitation laser over the various accessible rovibronic transitions from the ground state, a range of rovibrational levels of the molecule may be probed. LIF is widely used for detection of small molecules such as OH and NO, and when

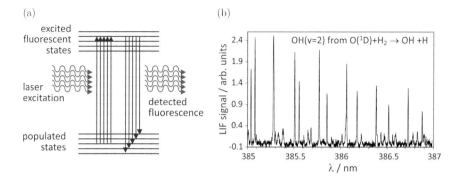

Fig. 6.5 (a) In LIF, molecules undergo laser excitation to a fluorescent state, and the resulting fluorescence intensity is used to probe the original population of the lower-state energy levels; (b) LIF spectrum of OH($v = 2$) reaction products formed in the reaction $O(^1D) + H_2 \rightarrow OH + H$.

coupled with a laser pump-probe scheme,[3] it can even be used to record quantum state populations of nascent reaction products. As an example, the spectrum shown on the right of Figure 6.5 is a LIF spectrum for OH molecules newly formed in the insertion reaction $O(^1D) + H_2 \rightarrow OH + H$.

6.1.4 Resonance-enhanced multiphoton ionization (REMPI)

A second technique that satisfies the sensitivity requirements for molecular beam spectroscopy is REMPI. Selective ionisation of the molecule of interest makes detection relatively straightforward. Ions can be steered towards a detector by means of an appropriately tuned electric field, and particle multiplier detectors capable of detecting single ions have been widely available for several decades. Single-photon ionisation of most molecules is unfeasible, since their ionisation energies correspond to photons in the soft X-ray region, which are not easy to generate in most laboratories. Instead, ionisation is usually brought about through essentially simultaneous absorption of two or more photons, as shown in Figure 6.6.

---

[3]Laser pump-probe experiments are widely used when investigating the kinetics and dynamics of elementary chemical reactions. The first (or 'pump') laser pulse initiates reaction, usually through photolysis of a chemical bond, and after a short delay, the second ('probe') laser pulse probes one or more of the reaction products spectroscopically.

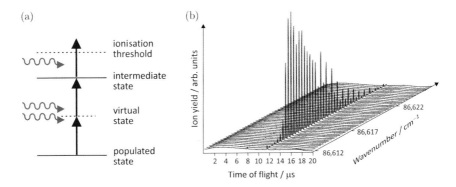

Fig. 6.6 (a) Schematic of a (2 + 1)REMPI transition, in which two photons are used to access an intermediate state, before a third photon ionises the molecule; (b) REMPI-TOF signal for jet-cooled HF.

Source: Adapted from Kvaran et al., J. Mol. Struct., **790**, 27–30 (2006).

The first photon accesses a 'virtual state'. Such a state is not a solution of the molecular Hamiltonian, but can be thought of as a very short-lived state of the 'molecule + photon' system. So long as a second photon arrives within a sufficiently short space of time — which requires high light intensities, usually from a laser pulse — the virtual state can absorb a photon to reach a real or virtual state of higher energy. Non-resonant multiphoton ionization, in which none of the intermediate levels accessed by successive photon absorptions correspond to real levels of the molecule, is not a particularly efficient process. However, when one of the intermediate states does correspond to an actual state of the molecule, the ionisation efficiency is increased enormously, often by orders of magnitude, yielding resonance-enhanced multiphoton ionization. REMPI schemes are usually labelled in terms of the number of photons required to reach the resonant state and the number of photons required to ionise from the resonant state. So, for example, in a (2 + 1)REMPI process, two photons are absorbed to take the molecule from the ground state to the intermediate state, and a third photon ionises the molecule from this state. The photon energy used for the initial transition to the resonant state can be tuned to ionise a specific quantum state of the molecule of interest, and scanning the laser wavelength across the rovibrational states of the molecule allows a REMPI absorption spectrum to be recorded.

The requirements for a molecule to exhibit a REMPI spectrum are considerably less stringent than those required for it to exhibit a LIF spectrum, and for this reason REMPI is more widely used than LIF. REMPI is often combined with time-of-flight mass spectrometry, in which ions of different masses are separated in time at the detector. This allows the signal from the molecule of interest to be isolated, even in the presence of other ions, e.g. ions produced from background gas in the vacuum chamber.

6.1.5 Cavity-enhanced absorption spectroscopy methods

In principle, absorption spectroscopy is a very appealing approach to determining spectra for virtually any chemical species. As shown in Figure 6.7(a), the method is experimentally extremely simple, requiring only a measurement of the intensity of light transmitted through the sample as a function of wavelength, as well as being universal: every atom or molecule absorbs light in some region of the electromagnetic spectrum. Unfortunately, a major shortcoming of the method is that it is often very insensitive. As outlined in Section 1.1.2, the measured absorption is determined by the Beer–Lambert law, reproduced below:

$$I(\lambda)/I_0(\lambda) = \exp(-n\sigma(\lambda)l) \tag{6.5}$$

where I and I_0 are the transmitted and incident light intensities, n is the number density and σ the absorption coefficient of the absorbing species, and l is the path length through the sample. The exponential dependence on number density means that when the number density is low, which is often

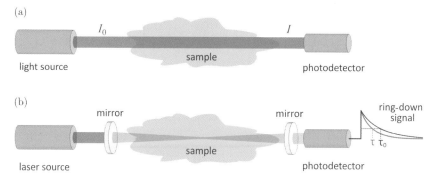

Fig. 6.7 (a) Schematic of a conventional absorption spectroscopy measurement; (b) Schematic of the experimental setup for a cavity ring-down or cavity-enhanced absorption measurement.

the case in laboratory experiments on unstable molecules of the type found in interstellar gas clouds, the ratio of transmitted to incident light intensities is very close to unity, and the sensitivity is very poor. When recording absorption spectra from interstellar space, the extremely low number densities are compensated to a considerable degree by the enormous path lengths involved, so that meaningful absorption spectra may still be measured. In the laboratory, the absorption path length may be increased artificially by modifying the experimental setup to include an optical cavity (see Figure 6.7(b)).[4] An optical cavity is a structure that traps light, the simplest example being a pair of carefully aligned mirrors. Incorporating an optical cavity into an absorption spectrometer can improve the detection sensitivity by many orders of magnitude, with detection limits in the parts per million to parts per billion range being achieved routinely.[5] A variety of CEAS methods have been developed, two of which will be considered in the following.

6.1.5.1 *Cavity ring-down spectroscopy*

The cavity ring-down technique was originally developed within the aerospace industry for characterising very highly reflective mirrors,[6] and was adapted for molecular absorption methods in the late 1980s by O'Keeffe and Deacon.[7] The technique is based on measuring the decay of light intensity within a high-finesse optical cavity in the presence and absence of the sample of interest. In the simplest configuration, shown in Figure 6.7(b), the cavity is formed from a pair of carefully aligned high-reflectivity planoconcave mirrors. A pulse of light, usually from a laser, is injected into the cavity along the axis through one of the mirrors, and the small amount of light coupled into the cavity undergoes repeated reflections between the two-mirrors. A small amount of light is lost on each reflection, primarily

---

[4]H. Linnartz, Cavity ring-down spectroscopy of molecular transients of astrophysical interest, Chapter 6 of *Cavity Ring-down Spectroscopy: Techniques and Applications*, eds. G. Berden and R. Engeln (John Wiley and Sons, 2009).

[5]G. Berden, R. Peeters, and G. Meijer, Cavity ring-down spectroscopy: Experimental schemes and applications, *Int. Rev. Phys. Chem.*, **19**(4), 565–607 (2000).

[6]J. M. Herbelin and J. McKay, Development of laser mirrors of very high reflectivity using the cavity-attenuated phase-shift method, *App. Opt.*, **20**(19), 3341–3344 (1981); M. Kwok, J. Herbelin, and R. Ueunten, Cavity phase–shift method for high reflectance measurements at mid-infrared wavelengths, *Opt. Eng.*, **21**(6), 979–982 (1982).

[7]A. O'Keeffe and D. A. G. Deacon, Cavity ring-down optical spectrometer for absorption measurements using pulsed laser sources, *Rev. Sci. Instrum.*, **59**, 2544–2555 (1988).

due to transmission or absorption by the mirrors or absorption or scattering by gas within the cavity. As a result, the light intensity within the cavity decays exponentially with time. The exponential decay may be recorded by placing a sensitive photodetector behind the second mirror to detect the small amount of light leaking out of the cavity through the mirror on each reflection. The time constant, τ, of the decay, also known as the 'ring-down time', is given by

$$\tau = \frac{nd}{c(L + \alpha Cl)} \quad (6.6)$$

where d is the cavity length, c/n the speed of light within the cavity (n being the refractive index of the medium inside the cavity), L the cavity loss per pass in the absence of sample (for a two-mirror cavity this is usually simply $1 - R$, where R is the mirror reflectivity), and α the absorption coefficient of an absorbing species present at concentration C over a path length l within the cavity.

Apart from instrumental factors such as the cavity length, intrinsic cavity losses, and the path length, the only parameters the ring-down time depends upon are the absorption coefficient and concentration of the sample. By recording the ring-down time in the presence and absence of the sample (denoted τ and τ_0, respectively), the absolute absorption per unit pathlength, $\kappa = \alpha C$, can be determined.

$$\kappa = \alpha C = \frac{nd}{cl}\left(\frac{1}{\tau} - \frac{1}{\tau_0}\right) \quad (6.7)$$

The limit of detection is determined by the minimum change in ring-down time, $\delta\tau_{\min}$ that can be determined reliably on introduction of a sample gas to the cavity. Conventionally, this is defined to be three times the standard deviation in the baseline ring-down time, τ_0. Making the approximation that at very low sample concentrations, $\tau\tau_0 \approx \tau_0^2$, by rearranging Equation (6.7), we find the minimum detectable absorption per unit path length to be

$$\kappa_{\min} = \frac{nd\delta\tau_{\min}}{cl\tau_0^2} \quad (6.8)$$

We see that optimising the sensitivity of a ring-down measurement relies primarily on minimising the baseline cavity losses in order to maximise τ_0. Many different schemes have been devised in order to achieve this.

In addition to the greatly enhanced path length, and therefore greatly enhanced sensitivity, cavity ring-down spectroscopy (CRDS) has an additional advantage over conventional absorption measurements. Because the time-dependence of the light intensity is measured, rather than the light intensity itself, the technique is largely insensitive to noise caused by shot-to-shot fluctuations in the intensity of the pulsed laser source. This tends to increase the sensitivity of the measurement even further.

6.1.5.2 *Cavity-enhanced absorption spectroscopy*

Cavity-enhanced absorption spectroscopy (CEAS), also known as integrated cavity output spectroscopy (ICOS), is an experimentally simpler variant of CRDS which does not require a pulsed laser source or a time-resolved measurement of the cavity output. In CEAS, a continuous-wave (or pseudo-continuous-wave) light beam is coupled into the cavity, and the intensity coupled out from the cavity is measured. The intensity transmitted through the cavity is proportional to the ring-down time, and therefore provides the necessary information on absorption without requiring a measurement of the decay constant. Since in CEAS we only measure a signal that is *proportional* to τ, rather than measuring τ itself, a calibration measurement is required in order to obtain absolute (rather than relative) absorption data. The calibration consists either of measuring the mirror reflectivity in a separate cavity ring-down measurement, or measuring the CEAS signal for a known sample of known number density.

The absorption per unit pathlength for a CEAS measurement is given to a very good approximation[8] by

$$\kappa = \frac{1}{d}\left(\frac{I_0}{I} - 1\right)(1 - R) \tag{6.9}$$

where d is the cavity length and R the mirror reflectivity, as before, and I_0 and I are the transmitted intensities in the absence and presence of an absorbing sample. A CEAS measurement is more sensitive by a factor of $(1-R)^{-1}$ than a single pass absorption measurement, and this factor is generally referred to as the *cavity-enhancement factor*.

Making the approximation that $(I_0 - I)/I \approx (I_0 - I)/I_0$, the minimum detectable absorption per unit pathlength is given by,

$$\kappa_{\min} = \frac{1}{d}\frac{\delta I_{\min}}{I_0}(1-R) \tag{6.10}$$

---

[8]S. E. Fiedler, A. Hese, and A. A. Ruth, Incoherent broad-band cavity-enhanced absorption spectroscopy, *Chem. Phys. Lett.*, **371**(3–4), 284 (2003).

where $\delta I_{\min}$ is the minimum detectable change in light intensity transmitted through the cavity.

Though CEAS does not provide absolute absorption measurements without a separate calibration step, for certain applications it has a number of advantages over CRDS. Firstly, the dynamic range is larger for CEAS. In CRDS, intense absorption signals yield ring-down times that are too short to determine with sufficient accuracy, whereas such signals pose no particular problems for a CEAS measurement. Additionally, broadband measurements are relatively straightforward in CEAS, simply requiring a broadband light source and spectral resolution of the cavity output. In contrast, such measurements are extremely difficult to carry out using CRDS, as the cavity output must be both spectrally and temporally resolved.

6.1.6 Molecular size considerations

We noted earlier that the accurate calculation of molecular energy levels becomes increasingly challenging as molecular size increases. Experimental spectroscopic studies on large molecules are also more challenging than on small molecules, though for different reasons. Firstly, in order to record a gas-phase spectrum of a molecule, it must be possible to prepare it in the gas phase. Any molecule with an appreciable vapour pressure is easily prepared in a supersonic expansion. However, more sophisticated methods must be used for molecules with minimal vapour pressure.

One approach is simply to heat the gas inside the molecular beam source. In some cases this is successful; however, particularly for organic species, heating often leads to decomposition or other undesirable chemical reactions. A relatively simple approach is to design a laser ablation source, in which the molecule of interest is prepared within a solid sample placed immediately in front of a molecular beam source. Illuminating the surface with a laser pulse of an appropriate energy produces a plume of molecules from the surface, some of which will be entrained within the molecular beam and cooled in the expansion.

A variety of methods have been developed by the mass spectrometry community to prepare gas-phase samples of involatile molecules,[9] usually in ionic form. These include electrospray methods, matrix-assisted laser

---

[9]G. Siuzdak, An introduction to mass spectrometry ionization, in *The Expanding Role of Mass Spectrometry in Biotechnology*, 2nd edn. (MCC Press, 2005).

desorption ionization (MALDI), laser-induced acoustic desorption (LIAD), desorption ionization on silicon (DIOS), and others. Such methods have considerable potential to be developed for use in gas-phase optical spectroscopy measurements, but at present the achievable number densities are rather too low for them to be useful.

An alternative to preparing molecules in the gas phase is to prepare them in a solid matrix of an inert gas at low temperature. The assumption in this approach is that if the concentration of the molecule of interest within the matrix is low enough, its interactions with the surrounding inert gas atoms will be fairly negligible, and the spectrum will therefore be similar to that of an isolated molecule in the gas phase. This may well be a reasonable approximation for electronic transitions, but molecular vibrations are likely to be significantly hindered in such an environment, and molecular rotations will be non-existent.

Assuming we can prepare a suitable sample of gas-phase molecules, there are further challenges associated with increasing molecular size. As the size of the molecule increases, so too does its density of rotational states. This not only leads to a dramatic increase in the complexity of the spectrum, but also to a reduction in signal levels due to the reduced population within any single rotational level. Some authors have even speculated that the severity of these challenges is such that they impose a fundamental limit on the size of molecule that can be detected in the interstellar medium through detection of its rotational emission.

6.2 Experimental astrochemistry II: Gas-phase kinetic and dynamical data

As noted in Section 5.2, the determination of accurate rate constants for all relevant chemical reactions is vital if realistic chemical models of molecular clouds are to be developed. Consequently, a number of experimental methods have been developed to measure rate constants for both ion–molecule reactions and neutral reactions of relevance to the interstellar medium. In the following sections, we will outline the basic principles underlying such measurements. The primary methods used to measure ion–molecule reaction rates to date are ion cyclotron resonance mass spectrometry (see Section 6.2.1), flow tube methods (Sections 6.2.2 and 6.2.3), and the CRESU method (the acronym, along with key features of the technique, will be explained in Section 6.2.4). In all of these approaches, the reactant temperatures tend to be much higher than the temperatures relevant to the interstellar medium, but the fact that the rates of ion–molecule reactions are relatively independent of temperature means that this is less of a problem than might initially be anticipated. In addition to more conventional methods, we will also look at a new method employing ions trapped in 'Coulomb crystals' (Section 6.2.5), which allows extremely low temperatures ($<1\,\text{K}$) to be accessed. Neutral reactions tend to be studied either via flow tube methods or with pulsed photolysis/laser pump-probe methods (see Section 6.2.4).

Rate constants have been measured for many chemical reactions of astrochemical interest. A detailed compilation of values can be found in a technical report[10] from NASA's Jet Propulsion Laboratory.

6.2.1 Ion cyclotron resonance mass spectrometry

6.2.1.1 *The ion cyclotron resonance technique*

One of the earliest methods used to measure rate constants for ion–molecule reactions was ion cyclotron resonance mass spectrometry (ICR-MS).[11] While this is a complex technique to understand in detail, the basic principles are fairly straightforward. An ICR-MS instrument consists of an ion source, an ion trap positioned within a high-field magnet, and ion optics to guide ions from the source region into the trap. The key features of the ion trap are shown in Figure 6.8, and comprise the strong uniform magnetic field $\mathbf{B}$ from the magnet, directed along the z-axis, together with three pairs of field plates arranged orthogonally along the x-, y-, and z-axes, known as the excitation, detection, and trapping plates, respectively. The reasons behind this naming convention will become apparent shortly.

Ions of charge q and mass m injected into the ion trap with velocity $\mathbf{v}$ are subject to a force $\mathbf{F}$ (known as the *Lorentz force*) given by

$$\mathbf{F} = m\frac{d\mathbf{v}}{dt} = q\mathbf{v} \times \mathbf{B} \tag{6.11}$$

---

[10] V. G. Anicich, An index of the literature for bimolecular gas phase cation-molecule reaction kinetics, Technical Report JPL-Publ-03-19, Jet Propulsion Laboratory, California Institute of Technology, Pasadena, CA, USA, November 2003.

[11] The first ICR-MS instrument was built by Comisarow and Marshall in the early 1970s, and is described in M. B. Comisarow and A. G. Marshall, Fourier transform ion cyclotron resonance spectroscopy, *Chem. Phys. Lett.*, **25**, 282 (1974). A good introduction to the technique is provided in A. G. Marshall, C. L. Hendrickson, and G. S. Jackson, Fourier transform ion cyclotron resonance mass spectrometry: A primer, *Mass Spectrom. Rev.*, **17**, 1–35 (1998).

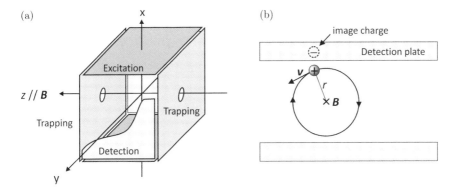

Fig. 6.8 (a) Schematic of an ion cyclotron resonance mass spectrometer, showing the excitation, detection, and trapping plates; (b) principle of detection within an ICR mass spectrometer. See text for details.

The vector cross product appearing in Equation (6.11) results in a force perpendicular to the plane defined by the ion velocity vector and the magnetic field vector. Assuming that the ion undergoes no collisions, and therefore maintains a constant speed, the result is that the magnetic field bends the ion trajectory into a circle of radius r. The resulting circular motion is known as *cyclotron motion*. To simplify the situation further, we will assume that the ion velocity lies in the xy plane, perpendicular to $\mathbf{B}$, and therefore has no component along the direction of the magnetic field. In practice, this is achieved by applying a trapping potential to the trapping plates in order to confine the ions along the z-axis within the ion trap. The magnitude of the force acting on the ion is then

$$F = m\frac{dv}{dt} = qvB\sin\theta = qvB \tag{6.12}$$

where θ is the angle between $\mathbf{v}$ and $\mathbf{B}$, in this case 90°. The linear and angular velocities, v and ω, of the ions are related by $\omega = v/r$, and for uniform circular motion, the acceleration dv/dt is equal to v^2/r. This allows us to simplify Equation (6.12) as follows:

$$m\frac{v^2}{r} = qvB$$
$$m\frac{v}{r} = qB$$
$$m\omega = qB \tag{6.13}$$

Rearranging the above and using the relationship between the angular velocity and the frequency, $\nu = \omega/(2\pi)$, yields an expression for the cyclotron frequency, ν_c

$$\nu_c = \frac{qB}{2\pi m} \qquad (6.14)$$

Note that all ions of a given mass-to-charge ratio, m/q, have the same cyclotron frequency, independent of their velocity. At typical field strengths of around 7 Tesla, ICR frequencies lie in the range from a few kiloHertz to a few megaHertz. Rearranging the second line of Equation (6.13) yields an expression for the radius of the cyclotron motion, which does depend on the ion velocity, as well as on the mass-to-charge ratio m/q.

$$r = \frac{mv}{qB} \qquad (6.15)$$

Thermal ions have cyclotron radii of only around 0.08 mm. In order to detect the ions, they must be excited to orbits of much larger radius. One way in which this may be achieved is by exposing the ions to an electric field oscillating in the plane of their cyclotron motion at a frequency equal to or close to the cyclotron frequency. The field is generated by applying equal and opposite sinusoidally oscillating potentials to the excitation plates positioned along the y-axis (see Figure 6.8). A reasonably straightforward vector treatment of the field[12] leads to the result that the resulting orbital radius depends on the amplitude of the electric field, E_0, and the time t for which it is applied

$$r = \frac{E_0 t}{2B} \qquad (6.16)$$

Note that Equation (6.16) is independent of mass, which means that all ions within a given range of m/q can be excited to the same orbital radius simultaneously. Exciting all of the ions simultaneously also means that their motion is spatially coherent, i.e. there is a well-defined phase relationship between ions of different mass-to-charge ratio, which have different cyclotron frequencies.

Detection of the ions occurs via the image charge they induce in the detection plates. As the ions undergo coherent cyclotron motion, they induce image charges as they pass each detection plate. The different image charges generated on the top and bottom detection plates cause

---

[12] A. G. Marshall, C. L. Hendrickson, and G. S. Jackson, Fourier-transform ion cyclotron resonance mass spectrometry: A primer, *Mass Spectrom. Rev.*, **17**, 1–35 (1998).

a time-varying current to flow, which constitutes the signal. Fourier transforming the data transforms the signal from the time domain to the frequency domain, yielding a spectrum of cyclotron frequencies which can be further transformed into a mass spectrum using Equation (6.14).

6.2.1.2 Measuring ion–molecule rate constants via ICR-MS

In the absence of collisions, ions of a given mass-to-charge ratio will continue orbiting indefinitely within an ICR mass spectrometer, yielding a continuous sinusoidal contribution to the time domain signal. If a neutral gas is admitted to the ion trap, the ensuing ion–molecule reactions convert reactant ions into product ions, leading to a damping of the time-dependent reactant ion signal. This is shown schematically in Figure 6.9. The damping of the signal can be modelled in terms of the kinetics of the ion–molecule reaction, allowing the rate constant to be determined. For example, if the neutral reactant is present in excess, such that the rate equation is pseudo-first-order, then the reactant ion signal will decay exponentially.

In addition to monitoring the decay of reactant ions, ICR-MS can also be used to monitor (simultaneously) the formation of product ions. However, the method does have some shortcomings. The neutral co-reactant must be a stable species, so ion-radical reactions cannot be studied. Also, the low pressures inside the mass spectrometer mean that the effects of third-body collisions are not observed. Such collisions are not important in the interstellar medium, but as noted in Section 4.6.1, studying collisionally-stabilised reactions in the laboratory can provide valuable information on their radiatively-stabilised analogues. Finally, ICR mass spectrometers are not low-cost instruments.

Fig. 6.9 Damping of the ICR signal with time due to reactive ion–molecule collisions within the ICR cell.

6.2.2 The flowing afterglow technique

Many early (pre-1980) measurements of ion–molecule reaction rates employed the flowing afterglow technique.[13] A typical instrument is shown schematically in Figure 6.10(a). Ions produced in the ion source are flowed in a helium buffer gas along a flow tube, typically around 1 m long and 80 mm in diameter, at a total pressure of around 0.5 Torr (67 Pa). Neutral co-reactants can be introduced at several points along the flow tube, and the reaction mixture is detected mass spectrometrically at the end of the tube. Rate constants are measured by recording the change in intensity of a reactant or product mass peak as the concentration of neutral reactant is varied. Extracting quantitative rate constants from the data requires

(a) Flowing afterglow

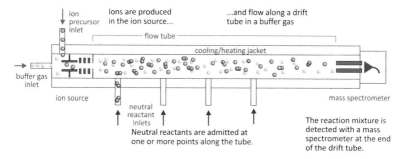

(b) Selected-ion flow tube

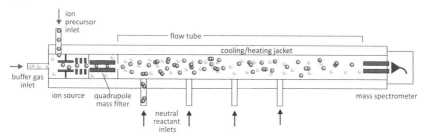

Fig. 6.10 Schematic of (a) a flowing afterglow instrument and (b) a selected-ion flow tube (SIFT). See text for explanation.

---

[13] For a history of the technique, see E. E. Ferguson, A personal history of the early development of the flowing afterglow technique for ion molecule reaction studies, *J. Am. Soc. Mass Spectrom.*, **3(5)**, 479–486 (1992).

a comprehensive understanding of the fluid flow dynamics within the flow tube.

Enclosing the flow tube in a jacket, which can be heated or cooled either electrically or by flowing through a liquid of the appropriate temperature, allows the reaction mixture to be heated or cooled in order to make measurements on the temperature dependence of the reaction rate.

Note that the gas discharge often leads to the formation of several different types of ion, so that there may be a number of competing ion–molecule reactions taking place within the flow tube, complicating the data analysis. This problem was solved by the introduction of the selected-ion flow tube, described in the following section.

6.2.3 The selected-ion flow tube

The simple flowing afterglow method outlined above has largely been replaced by a somewhat more sophisticated approach known as the selected-ion flow tube (SIFT) method,[14] in which a quadrupole mass filter positioned in between the ion source and the flow tube is used to select a single ion of interest for injection into the flow tube. In other respects, the method is essentially the same as the flowing afterglow method described in the previous section. A schematic of a SIFT is shown in Figure 6.10(b).

6.2.4 The CRESU method

CRESU stands for 'Cinétique de Réaction en Ecoulement Supersonique Uniforme', which translates as 'Reaction Kinetics in Uniform Supersonic Flows'.[15] In contrast to the flowing afterglow and SIFT techniques, CRESU allows rate constants to be measured for ion–molecule reactions at temperatures comparable with those in interstellar gas clouds. The technique is based on expanding a high-pressure gas through a specially shaped nozzle called a de Laval nozzle (see Figure 6.11). The nozzle is designed to collimate the gas into a uniform, collision-free supersonic beam at a characteristic temperature, and by adjusting the nozzle design, gas flows with temperatures ranging from room temperature down to around 10 K may be

---

[14] N. G. Adams and D. Smith, The selected-ion flow tube (SIFT); a technique for studying ion-neutral reactions, *Int. J. Mass Spectrom.*, **21**(3–4), 349–359 (1976).
[15] I. R. Sims, Gas-phase reactions and energy transfer at very low temperatures, *Annu. Rev. Phys. Chem.*, **46**, 109–137 (1995).

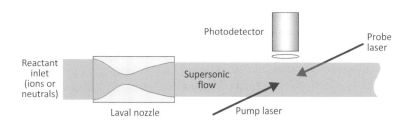

Fig. 6.11 Schematic of a CRESU instrument. See text for description.

achieved. The gas density in the expansion is relatively high (in the range from 10^{16} to 10^{18} cm^{-3}), so that thermal equilibrium is maintained at all times, and the flow is uniform with respect to temperature, density, and velocity. Note that this is in marked contrast to a pulsed molecular beam of the type described in Section 6.1.1, in which no thermal equilibrium, and therefore no well-defined temperature, is established.

The CRESU method can be used to study the kinetics of both ion–molecule and neutral reactions. Kinetics measurements within the gas flow are generally performed using laser pump-probe techniques (see footnote 3). The pump laser generates a reactive species through photolysis or ionisation of a suitable precursor molecule, and product detection in the probe step is often through LIF (see Section 6.1.3). Recording the signal as a function of pump-probe delay and/or reactant concentrations allows the rate constant to be determined.

6.2.5 Coulomb crystals

Experiments in which the ion is trapped in a Coulomb crystal also allow temperatures commensurate with those in the interstellar medium to be accessed. In order to understand the Coulomb crystal method, we first need to learn a little about laser cooling and trapping of atoms. Steve Chu, Claude Cohen-Tannoudji, and William Phillips were the first to produce ultracold atoms via laser cooling and trapping, and won the 1997 Nobel Prize in Physics for their efforts.

Laser cooling is based on transferring momentum from the photons in a laser beam to the atoms to be cooled, in order to slow the atoms down. Each time an atom absorbs a photon, a small amount of momentum is transferred to the atom, slightly changing its velocity. Many absorption-emission cycles are required in order to slow an atom down appreciably,

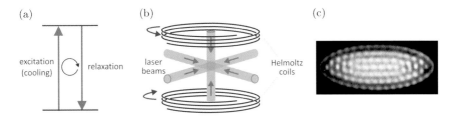

Fig. 6.12 (a) A closed optical loop is used for laser cooling of atoms. (b) A magneto-optical trap (MOT) is used for Döppler cooling and trapping of cooled atoms. (c) A Coulomb crystal of Rb atoms stored within a MOT.
Source: Data from S. Willitsch, M. T. Bell, A. D. Gingll, S. R. Procter, and T. P. Softley, Phys. Rev. Lett., **100**, 043203 (2008).

and so an atomic transition is required that forms a *closed optical loop*. This means that after absorption, spontaneous emission returns the atom back to its original state, as shown in Figure 6.12(a). Atoms with suitable two-level closed systems include alkali metals, metastable rare gas atoms, and singly charged alkali-earth ions. Atoms that emit to other states can also be laser cooled, but an additional laser is required in order to excite a suitable transition to return the 'lost' atoms to the optical loop.

The most common laser cooling technique is called Döppler cooling, and usually involves an experimental setup consisting of six intersecting laser beams, as shown in Figure 6.12(b). This is usually achieved using three orthogonally propagating beams, each reflected back along its own beam path by a mirror. The frequency of each laser beam is tuned slightly to the red of an atomic transition. Due to the Döppler effect, atoms moving towards the light source absorb more photons than those moving away, and are slowed. The excited state will of course receive a second momentum 'kick' when it emits a photon to return to the ground state, but the direction of this second kick is random, so that the net force after many transitions is along the laser beam direction. The three-axis laser beam configuration means that atoms are slowed down no matter which direction they are moving in. Temperatures as low as 150 μK have been achieved for ^{85}Rb using this technique.

To prevent cooled atoms from falling out of the laser interaction region under the force of gravity, Döppler cooling is often combined with a magnetic trapping force, yielding a magneto-optical trap (MOT), as shown in Figure 6.12(b). Zeeman splitting of the energy levels in the magnetic field increases as atoms move away from the centre of the trap, shifting the atomic resonance closer to the laser frequency, and increasing the chance of the atom receiving a photon 'kick' towards the centre of the trap when the appropriate (circular) polarisation of light is employed.

When ions are laser cooled and trapped, they form a crystalline array in which their positions are approximately fixed (with some small degree of motion caused by the oscillating trap potentials). The relative positions of the ions are determined by the balance between the confining forces of the trap and the Coulomb repulsion of the ions, with typical spacings on the order of 10–20 μm. During the laser cooling process, the ions are continuously absorbing and emitting photons, and by detecting the emitted fluorescence, an image of the crystal may be obtained. Such an image is shown in Figure 6.12(c).

Coulomb crystals may be stored for several hours in a suitable vacuum, and provide an ideal method for studying ultracold collisions between ions and other chemical species for cases in which a suitable cooling scheme can be identified for the ion.[16] When an individual ion reacts to form another chemical species, it is no longer resonant with the cooling laser, and its fluorescence is extinguished. Reaction rates can therefore be studied by introducing a low pressure of a neutral reactant gas into the MOT and measuring the rate at which the ion signals disappear one by one. For example, Softley et al.[17] have combined Coulomb crystals with a velocity-selected molecular beam to study the kinetics of the reaction $Ca^+ + CH_3F \rightarrow CaF^+ + CH_3$ at relative translational temperatures down to 1 K.

6.2.6 Neutral reactions

The kinetics of neutral reactions can be studied using flow tube methods similar in many ways to the flowing afterglow technique described in Section 6.2.2. Radical species are formed at the entrance to a Pyrex flow tube, and reactants or products are detected optically or mass spectrometrically at the far end of the flow tube. The inner wall of the flow tube is usually treated with wax or Teflon to minimise reactions of the radicals with the wall. Radicals such as atomic hydrogen, nitrogen, and oxygen can be formed via dissociation of the corresponding diatomic in a microwave

---

[16] M. Drewsen, I. Jensen, J. Lindballe, N. Nissen, R. Martinussen, A. Mortensen, P. Staanum, and D. Voigt, Ion Coulomb crystals: A tool for studying ion processes, Int. J. Mass Spectrom., **229**(1–2), 83–91 (2003).

[17] S. Willitsch, M. T. Bell, A. D. Gingell, S. R. Proctor, and T. P. Softley, Phys. Rev. Lett., **100**, 043203 (2008).

discharge, carbon atoms can be formed by reacting CCl_4 or CBr_4 with an alkali metal, and molecular radicals can be formed by reacting atoms with a suitable co-reactant. Rate constants may be extracted either by varying the concentration of one of the reactants within the flow tube (i.e. monitoring the reaction mixture at a fixed time, determined by the flow velocity and flow tube length, as a function of reactant concentration), or by varying the flow rate (thereby monitoring the reaction mixture as a function of reaction time). Alternatively, neutral reaction rates can be measured via laser pump-probe methods. The first laser produces the radical species of interest via photolysis of a suitable precursor in the presence of a much higher concentration of the second reactant. The reaction then occurs under pseudo-first-order conditions. The first-order decay in the concentration of the radical thus produced is then followed using a suitable spectroscopic technique, often LIF, and the rate constant is simply the time constant of the resulting exponential decay.

6.3 Experimental astrochemistry III: Dust-grain chemistry

In Section 4.6.3, we introduced chemistry occurring on the surface of interstellar dust grains. These grains consist of a silicate or carbonate core surrounded by layers of ice formed by condensation of molecules onto the core from the surrounding molecular gas. As we have come to appreciate the importance of dust grain chemistry in molecular synthesis within the interstellar medium, there have been an increasing number of laboratory studies into both the formation and structure of interstellar ices and the chemical reactions occurring on their surfaces.

Forming a model interstellar ice in the laboratory usually starts by preparing a clean graphite surface under vacuum and cryogenically cooling it to temperatures of 5–20 K. Ices are then formed by exposing the surface to the gas of interest, e.g. H_2O, CH_3OH, or NH_3. Alternatively, the ice can be deposited on a transparent window. The latter approach is often employed when the ice is to be studied spectroscopically.

6.3.1 Ice structures via infrared spectroscopy

The structure of the resulting ice is usually probed via infrared (IR) spectroscopy, carried out either in transmission mode or reflection mode. Both modes are illustrated in Figures 6.13(a) and 6.13(b). If the ice is

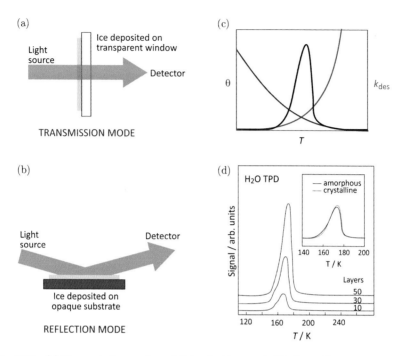

Fig. 6.13 (a) Transmission-mode and (b) reflection-mode IR absorption measurements on laboratory ice samples; (c) The peak in a temperature-programmed desorption measurement results from competition between the increase in desorption rate constant with temperature and the decrease in the number of adsorbed molecules; (d) Sample set of data for temperature-programmed desorption of water ices from an alumina surface, recorded as part of a study by Tzvetkov *et al.* on the interaction of glycine with ice nanolayers (see main text for reference).

deposited on a window, an IR absorption spectrum can be acquired in transmission mode simply by monitoring the intensity of the IR radiation transmitted through the sample and using the Beer–Lambert law (see Section 1.1.2) to analyse the data. Transmission-mode measurements are sensitive to the bulk structure of the ice. However, the catalytic properties of interstellar ices are determined largely by the structure at the surface. This can be probed by a standard surface-science technique known as reflection-absorption infrared spectroscopy (RAIRS). In RAIRS, an IR beam is directed towards the surface at a shallow angle of incidence, and the spectrum of the reflected beam is analysed. The shallow angle of incidence has the consequenece that RAIRS is sensitive only to the surface layers of the ice, and not to the bulk structure. Excitation of normal modes of adsorbed molecules, or excitation of adsorbate-surface bonds, leads to depletion of intensity in the reflected beam at the corresponding characteristic frequencies.

Ices are found to be largely amorphous, with differing degrees of porosity at the surface.

6.3.2 Thermodynamics of adsorption and desorption via temperature-programmed desorption

The energetics of adsorption and desorption of molecules onto the ice surface are usually studied using another surface-science technique known as temperature programmed desorption (TPD). The adsorbate(s) to be studied are deposited on the surface, and the surface is then heated at a constant rate of around 2–10 K s^{-1}. A mass spectrometer is used to record the number of molecules desorbed from the surface as a function of time (and therefore surface temperature) during the temperature sweep.

Desorption is an activated process, and the desorption rate constant therefore follows an Arrhenius dependence. However, the overall rate also depends on the number of molecules remaining on the surface, which decreases as the surface temperature is increased. The result is that the TPD signal for a given molecule, i.e. a plot of surface coverage θ vs temperature T, takes the form of a peak, which can be described by the Polanyi–Wigner equation.

$$-\frac{d\theta}{dT} = \frac{\nu \theta^n}{\beta} \exp\left(-\frac{E_{\text{des}}}{RT}\right) \quad (6.17)$$

where β is the heating rate (in K s^{-1}), n is the kinetic order of the dissociation, E_{des} is the desorption energy, and R is the gas constant. The parameter ν is a frequency parameter, which can be determined along with E_{des} by fitting the experimental data to Equation (6.17). In simple cases, the frequency parameter can be interpreted as the vibrational frequency of the adsorbate-surface bond.

A sample set of data for TPD of water ices from an alumina surface is shown in Figure 6.13(d). The data were recorded as part of a study by Tzvetkov and coworkers into the interaction of glycine with ice nanolayers.[18] The main plot shows TPD spectra for 10, 30, and 50-layer low density

---

[18] G. Tzvetkov, M. G. Ramsey, and F. P. Netzer, Interaction of glycine with ice nanolayers, *Chem. Phys. Lett.*, **397**(4–6), 392–396 (2004).

amorphous (LDA) ice films deposited on Al_2O_3 at 110 K, and the inset compares the TPD spectra for a 50-layer LDA ice film and a 50-layer polycrystalline film. Slight differences between the two spectra indicate that the binding energy of water to the surface depends on the structural properties of the ice.

6.3.3 Photoinitiated molecular synthesis in interstellar ice analogues

Irradiation of dust grains by UV and vacuum ultraviolet (VUV) photons and/or cosmic ray impacts with the surface can initiate a considerable amount of chemistry through the generation of reactive species both on the surface and in the bulk of ice grains. Molecular synthesis of this nature has also been investigated in the laboratory. Interstellar ice analogues are generated as described above, and the surface is then irradiated with UV light. This is followed by a TPD experiment to quantify any new species generated. UV photoprocessing of ices consisting of simple molecules such as H_2O, CO, NH_3, and CH_3OH has been shown to generate a variety of new species, including simple species such as H_2, H_2CO, CO_2, CH_4, and HCO, as well as more complex molecules such as ethanol, ketones, amides, hexamethylenetetramine (HMT), and polyoxymethylene-related species.[19]

6.3.4 Formation of H_2 on ice surfaces

One of the most important processes to occur on the surface of interstellar dust grains is the synthesis of molecular hydrogen from two adsorbed H atoms. This process has been studied in detail in a number of different laboratories[20] by bombarding a variety of surfaces with atomic hydrogen and making spectroscopic measurements on the resulting H_2. It is clear that H_2 is indeed formed on ice surfaces, with the dynamics of the process depending on the surface structure. On non-porous surfaces, the process is fast and direct, and H_2 molecules are released immediately from the surface with a considerable degree of vibrational excitation. On porous surfaces, the newly-formed H_2 tends to thermalise, presumably becoming trapped in the

---

[19]L. J. Allamandola, M. P. Bernstein, S. A. Sandford, and R. L. Walker, Evolution of interstellar ices, *Space Sci. Rev.*, **90**, 219–232 (1999).

[20]D. A. Williams, W. A. Brown, S. D. Price, J. M. C. Rawlings, and S. Viti, Molecules, ices, and astronomy, *Astron. Geophys.*, **48**, 1.25–1.34 (2007).

pores and undergoing multiple collisions on or with the surface before being released into the gas phase.

6.4 Case study: Ethylene glycol

Now that we have provided an overview of the various theoretical and experimental methods at our disposal for identifying and studying interstellar molecules, we will finish our discussion of laboratory-based astrochemistry with a brief case study, in which observational and laboratory data are brought together in order to identify a newly-detected molecule in interstellar space, and to gain some insight into the processes leading to its formation. The molecule in question is ethylene glycol, which has the structure HO–CH$_2$–CH$_2$–OH. Ethylene glycol is probably best known on Earth for its widespread use in antifreeze products for the automobile industry. It is also a key organic molecule thought to be associated with prebiotic sugar synthesis. The lowest-energy conformation is shown in Figure 6.14.

Ethylene glycol in the interstellar medium was first reported by Hollis et al.[21] following the detection of a series of millimetre-wave emissions originating from the molecular cloud Sagittarius B2. The transitions arise from torsional motion of the two OH groups, and had been studied previously in the laboratory by Christen et al.[22] allowing them to be assigned unambiguously.

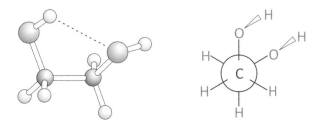

Fig. 6.14   Lowest-energy conformation of ethylene glycol.

Source: Adapted from Hollis *et al.* (2002).

---

[21] J. M. Hollis, F. J. Lovas, P. R. Jewell, and L. H. Coudert, Interstellar antifreeze: Ethylene glycol, *Astrophys. J.*, **571**, L59–62 (2002).
[22] D. Christen, L. H. Coudert, R. D. Suenram and F. J. Lovas, The rotational/concerted torsional spectrum of the g'Ga conformer of ethylene glycol, *J. Mol. Spectrosc.*, **172**, 57–77 (1995).

As an asymmetric-top molecule, the rotational energy level structure of ethylene glycol is complex. To complicate matters further, the various possible torsional motions mean that tunnelling splittings[22] must also be considered, as must further Coriolis (rovibronic or rotation-vibration) couplings between the tunnelling motion and the overall rotation of the molecule. In order to assign the transitions measured in the laboratory, a theoretical model was developed which treated the rovibrational energy as the sum of a rigid rotor energy (as described in Section 5.3.1), and a tunnelling splitting. Each level was then perturbed by the appropriate Coriolis coupling. The steps required to set up the molecular Hamiltonian based on this formalism, and to solve it to determine the energy levels, are fairly involved, and we shall not concern ourselves with them here (interested readers can read about this process in Christen's original paper). The diagram in

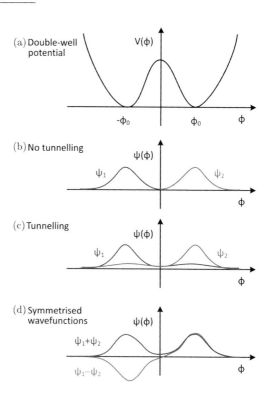

[23]In ethylene glycol, torsional motion of the two OH groups leads to a molecular potential containing two minima separated by a barrier, as shown in panel (a). Whenever such a potential arises, quantum mechanical tunnelling through the barrier becomes possible. Rather than being localised in one or other of the minima (panel (b)), the wavefunction for the system becomes spread over both minima, as shown in panel (c) for the ground-state vibrational wavefunctions of the system. The wavefunctions $\psi(\theta)$ must also satisfy the requirement that they have a definite symmetry, i.e. $\psi(\theta) = \pm\psi(-\theta)$. This is achieved by taking sums and differences of the wavefunctions in panel (c), as shown in panel (d). The resulting correctly symmetrised wavefunctions shown in panel (d) have slightly different energies; hence the presence of the barrier has caused a *tunnelling splitting* of the vibrational energy levels.

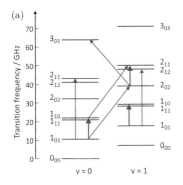 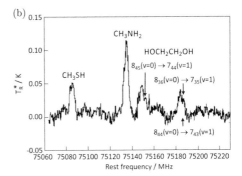

Fig. 6.15 (a) Predicted energy levels and transitions for ethylene glycol, taken from Christen et al.; (b) a section of the interstellar ethylene glycol spectrum recorded by Hollis et al.

Note: See footnotes for full references.

Table 6.1 Table of data for ethylene glycol.

$\nu/$ MHz	$T_R^*/$ mK	$\Delta V/$ km s^{-1}	$V_{\text{LSR}}/$ km s^{-1}	Obs. RI	Calc RI (50 K)	Calc RI (200 K)	$10^{-14} N$ (50 K)/ cm^{-2}	$10^{-14} N$ (200 K)/ cm^{-2}
75151.4	46	34	—	2.00	0.58	0.70	<14.5	<77.8
75186.1 (2 lines)	33	24	71.1	1.43	1.31	1.52	3.3	17.9
75299.9	23	25	70.7	1.00	1.00	1.00	3.1	19.8
92975.9	38	40	—	1.65	1.38	1.56	<5.9	<33.5

Note: See text for explanation of parameters.

Figure 6.15(a), taken from Christen's paper, shows some of the energy levels and transitions predicted by the model. The quantum number v relates to the tunnelling motion associated with concerted torsional motion within the molecule, while the other quantum numbers labelling the states correspond to rotational motion about the various molecular axes. Also shown, in Figure 6.15(b), is a section of the spectrum recorded from Sagittarius B2 by Hollis et al.

Table 6.1 shows some of the data reported by Hollis et al. based on their observations and the previous laboratory-based work by Christen et al. The table in the original publication contains considerably more data on each of the transitions than Table 6.1. Here we focus on the information that provides insight into the local environment of ethylene glycol in interstellar space.

The column headed T_R^* reports the 'brightness temperature', i.e. the temperature of a black body[24] with the same brightness as observed from the source. ΔV is a measure of the linewidth of the transition, and V_{LSR} is a measure of the Döppler shift (LSR stands for 'local standard of rest'). The Döppler shift of the entire Sagittarius B2 molecular cloud is around $64 \, \text{km s}^{-1}$. Values different from this reflect additional local motion, e.g. rotational motion of the cloud, which can help to localise the molecular source within the cloud. The remaining columns report relative intensities of the transitions, and compare these with the predicted relative intensities and number densities based on a Boltzmann distribution at two different temperatures. Comparison of these values indicates that the first transition is anomalously strong. This probably means that there is an overlapping contribution to this line from another molecular species. This is also consistent with the relatively broad width of the line relative to the 24–25 km s^{-1} linewidths of most of the other lines. The last line is also too broad to arise solely from an ethylene glycol transition, and probably contains an overlapping contribution from another molecule.

Ethylene glycol is a relatively large molecule, but its presence fits in with previous observations of aldehydes together with their corresponding reduced alcohols (glycolaldehyde has also been observed in the interstellar medium). Others in the series include formaldehyde/methanol, and acetaldehyde/ethanol. The mechanism(s) by which such molecules are synthesised in the interstellar medium have not been confirmed, but a number of laboratory studies have shed light on some possible pathways to their formation. Hiraoka *et al.*[25] have demonstrated that reaction between gas-phase atomic hydrogen and CO ice at 10–20 K produces both formaldehyde and methanol. In other experiments, Moore and Hudson[26] have synthesised alcohols and other complex molecules in the laboratory by subjecting samples of molecular ices to ionising radiation at a temperature of 15 K. They concluded that H and OH addition reactions occurring in the solid

---

[24] Black body radiation is a term used to describe the spectrum of radiation emitted from an object held at a constant, uniform temperature. The spectrum of emitted frequencies, ν, at temperature T is defined by Planck's law:

$$I(\nu, T) = \frac{2\nu^3}{c^2 (\exp(h\nu/k_B T) - 1)} \tag{6.18}$$

where h is Planck's constant, c is the speed of light, and k_B is Boltzmann's constant.
[25] K. Hiraoka, N. Ohashi, Y. Kihara, K. Hamamamoto, T. Sato, and A. Yamashita, *Chem. Phys. Lett.*, **229**, 408 (1994).
[26] Moore and Hudson, *Icarus*, **135**, 518 (1998).

phase may be the source of many of the molecular species observed in the interstellar medium. Such studies would imply that interstellar ethylene glycol is most probably formed either through surface-catalysed reactions on dust grains, or solid-state reactions occurring within the icy mantle of dust grains.

6.5 Summary

At this point, we have armed ourselves with an overview of the types of chemical processes occurring in interstellar space, beginning with the synthesis of chemical elements within stars and continuing with a summary of the various processes that lead to the formation of complex molecules within interstellar gas clouds. We have also considered a range of theoretical and experimental methods that may be employed to help identify molecules in space and to unravel the chemical processes by which they are formed. Furthering our understanding of the fascinating and rapidly evolving field of astrochemistry relies on success in interpreting spectra obtained from both astronomical and terrestrial measurements, in kinetic modelling of reaction cycles, and in laboratory measurements of rate constants and investigations into reaction mechanisms.

Having focused almost solely on gas-phase chemical and physical processes up until this point, we now turn to the condensed phase as we outline the processes leading to formation of the Solar System and planet Earth.

6.6 Questions

6.6.1 Essay-style questions

Q6.1 Explain why molecular beams are widely used in studies of astrochemically relevant molecules.

Q6.2 Provide a critical overview of the various spectroscopic techniques available for studying astrochemically relevant molecules. What types of information can each technique provide, and why might one technique be preferred over another?

Q6.3 Why does it become difficult to carry out spectroscopic studies for comparison with observational data as molecular size increases? How may some of the difficulties be overcome?

Q6.4 Most of the experimental methods used to measure the kinetics of reactions of relevance to astrochemistry employ temperatures much higher than those in the interstellar medium. Explain whether or not data from such measurements are useful.

Q6.5 Provide an overview of experimental methods available for studying the structure and chemical composition of interstellar ices.

6.6.2 Problems

P6.1 *Terminal velocity of a molecular beam*

Assuming that H_2 behaves as an ideal gas, with heat capacities $C_V = 3R/2$ and $C_p = 5R/2$, with R the gas constant, calculate the terminal velocity of a beam of H_2 molecules expanded from a source at room temperature. Repeat the calculation for CO_2.

P6.2 *CRDS of acetonitrile*

(a) Derive Equation (6.7) from Equation (6.6), showing that the absolute absorption per unit pathlength in a CDRS experiment can be determined from measurements of the ring-down time in the presence and absence of the sample.

(b) An empty cavity of length 30 cm has a ring-down time of 13.2 μs. Assuming it is possible to measure a 0.1% change in ring-down time, determine:

(i) the minimum detectable absorption per unit pathlength for a gaseous sample admitted to the cavity;

(ii) the minimum partial pressure of acetonitrile, CH_3CN, that can be detected via an IR transition with an absorption coefficient of 2×10^{-4} mbar^{-1} cm^{-1}.

P6.3 *Ion cyclotron resonance*

The reaction between C^+ ions and H_2 was studied inside an ion cyclotron resonance mass spectrometer with a field strength of 7 T.

(a) Calculate the cyclotron frequency, ν_c, for the C^+ ions inside the spectrometer.

(b) Calculate the radius r_c of the cyclotron motion for a sample of C^+ ions at 300 K. You may make the approximation that all ions have a velocity equal to the mean thermal velocity.

(c) The cyclotron radius following excitation by a radiofrequency pulse is given by:

$$r_c = \frac{E_0 t}{2B}$$

(i) Define the variables E_0 and t.
(ii) By writing out the dimensions of E_0, t and B in terms of SI base units and substituting back into the above expression, show that r_c has dimensions of metres.

(d) An RF pulse of amplitude 200 mV is applied to the ions in (b) for 2.8 ms to excite the ions to a higher-lying orbit. The excitation plates are separated by 1 cm. Calculate the new cyclotron radius.

(e) H_2 is now admitted to the spectrometer, leading to loss of C^+ ions through ion–molecule reaction with H_2. Assuming that H_2 is present in large excess relative to C^+, show that the concentration of C^+ ions (and therefore the ion cyclotron resonance signal) decays with time according to the expression

$$[C^+] = [C^+]_0 \exp(-k[H_2]_0 t)$$

where $[C^+]_0$ and $[H_2]_0$ are the initial concentrations of C^+ and H_2, and k is the bimolecular rate constant for the ion–molecule reaction.

P6.4 *Kinetics of the charge-transfer reaction $H^+ + NO \rightarrow H + NO^+$ via SIFT measurements*

The charge-transfer reaction $H^+ + NO \rightarrow H + NO^+$ has the following rate law, with rate constant k.

$$\frac{d[H^+]}{dt} = -k[H^+][NO]$$

(a) Show that if the reaction is studied under conditions in which NO is present in great excess over H^+, such that its concentration remains essentially constant over the course of the reaction, the rate law may be integrated to give

$$\ln[H^+] = -k[NO]t + \ln[H^+]_0$$

(b) In a SIFT experiment, the reaction time, t, may be determined from the flow velocity, v, and the 'reaction length' (i.e. the distance within the flow over which reaction occurs), z. Given a reaction length of 50 cm and a flow velocity of 90 ms^{-1}, determine the reaction time.

(c) The following data were recorded for the above reaction. Use the data, together with the integrated rate law above, to determine the rate constant for the H$^+$ + NO charge-transfer reaction.

[NO]/molecule cm^{-3}	[H+]/molecule cm^{-3}
0.00	3.70×10^6
5.00×10^{11}	1.89×10^4
1.00×10^{12}	96.37
1.50×10^{12}	0.49

P6.5 *Spectroscopic study of a diatomic molecule in Sagittarius B2*

This problem is based on data reported by D. A. Neufeld, J. Zmuidzina, P. Schilke, and T. G. Phillips, *Astrophys. J. Lett.*, **488**, L141 (1997).

A diatomic molecule in the giant molecular cloud Sagittarius B2 is detected via its $J = 2$ to $J = 1$ rotational transition at $121.6973\,\mu$m.

(a) The transition can only be observed from a space-based telescope. Suggest a reason for this.

(b) The transition wavelength quoted above has been corrected for the Döppler effect. The radial velocity of Sagittarius B2 relative to the Earth is $61.6\,\text{km s}^{-1}$. At what wavelength would the transition be detected?

(c) Ignoring centrifugal distortion, determine an expression for the transition energy in terms of the rotational constant and rotational quantum numbers of the upper and lower states, and hence calculate the rotational constant for the molecule.

(d) The true rotational constant for the molecule is $B_e = 20.9557\,\text{cm}^{-1}$. Explain the reasons for the discrepancy between this value and the value you calculated in part (c). Explain how additional experimental data could be used to determine a more accurate value for the rotational constant.

(e) Given that one of the atoms is hydrogen (i.e. the diatomic molecule has the formula HX), and that the bond length is 92 pm, determine the identity of the second atom, X, and therefore identify the molecule. Use the correct value for B_e given in part (d) in your calculations.

(f) The Döppler width $\delta\lambda$ of a spectral line due to thermal broadening is given by

$$\delta\lambda = \frac{2\lambda}{c}\left(\frac{2k_\mathrm{B}T\ln 2}{m}\right)^{1/2}$$

where k_B is Boltzmann's constant, T is the temperature, c is the speed of light, and m is the molecular mass. The measured linewidth is $1.5713 \times 10^{-4}\,\mu\mathrm{m}$. What is the temperature in the observed region of the cloud?

Chapter 7

Formation of the Solar System and the Evolution of Earth

In Chapter 3, we discovered that stars are formed through gravitational collapse within an interstellar gas cloud. In this chapter, we will revisit the process of gravitational collapse in more detail, and will find that it leads not only to the formation of stars, but also to a diverse range of other bodies, including planets of varying compositions, comets, and meteorites. Since we know most about the astronomical objects close to Earth, we will focus on the events leading to the formation of our own Solar System. Any model of these events must explain a number of known facts about our Solar System, namely:

(1) All of the planetary orbits lie in a single plane, with all planets orbiting the Sun in the same direction in near-circular orbits;

(2) The Sun and all planets (except Venus) all rotate about their axes in the same direction, generally with very small tilts (obliquities) between their orbital and equatorial planes (Neptune is the exception, with a large tilt);

(3) The planets are fairly regularly spaced from the Sun, and their composition varies with distance from the Sun. Dense, metal-rich planets are found close to the Sun, while giant gas planets occur further away;

(4) The Sun accounts for around 99.8% of the mass in the Solar System, but only around 0.2% of the angular momentum, i.e. the Sun rotates more slowly than we might expect;

(5) The pattern of planets in circular orbits in a plane around the Sun with a constant direction of rotation is mirrored by the pattern of moons in orbit around the planets;
(6) Comets are found in a large, spherical cloud surrounding the Solar System.

While we are still some way from a complete understanding of the formation of the Solar System, the *nebula* model is able to explain all of the above observations, and is the most widely accepted model at present.

7.1 The Solar nebula

The Solar System formed around 4.5 billion years ago during the gravitational collapse of part of a large, slowly swirling molecular gas cloud known as the Solar nebula. Such clouds can be stable for very long periods of time, with the gas pressure balancing the force of gravity. However, a small perturbation in the gas cloud, perhaps the result of a shock wave from a nearby supernova explosion, can be enough for gravity to tip the balance, initiating gravitational collapse. As the cloud collapses, conservation of angular momentum (see Section 5.4.9) causes its speed of rotation to increase. A further consequence of angular momentum conservation is that the cloud flattens into a disk shape (known as an *accretion disk* or *protoplanetary disk*) as the gravitational collapse continues. The details of accretion disk formation are beyond the scope of this text, but the basic physical principles are relatively easy to understand. Conservation of angular momentum in the plane perpendicular to the rotation axis will tend to resist gravitational collapse towards the axis of rotation (this effect can also be described in terms of 'centrifugal forces' pushing material outwards in this plane), while in directions parallel to the axis of rotation there is no such restriction, and material 'falls' into a disk rotating about the central protostar.

Of course, in order for the protostar to accrete sufficient matter to initiate fusion, there must be some mechanism for material in the accretion disk to move inwards towards the centre of the nebula. A particle of mass m and velocity v in an orbit of radius r from the newly forming protostar has angular momentum $L = mvr$. When the particle moves inwards under the force of gravity to an orbit of smaller radius, it loses gravitational energy and gains kinetic energy. Conservation of angular momentum would dictate

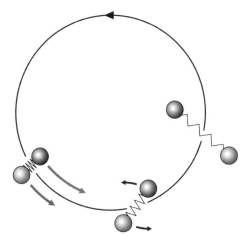

Fig. 7.1 Ball and spring analogy of angular momentum transfer in an accretion disk. The spring represents an attractive force (e.g. a magnetic force in some models). As the inner particle orbits faster than the outer particle, the spring is stretched. The restoring force slows down the inner particle and speeds up the outer particle, at the same time causing the inner particle to move to smaller orbital radius and the outer particle to move to larger orbital radius.

that the orbital velocity of the particle must increase if the orbital radius decreases. However, viscous and magnetic drag forces near the centre of the nebula mean that in fact the particle loses angular momentum. Since the overall angular momentum of the nebula must be conserved, the 'lost' momentum must be transferred to particles at larger radius. In this way matter is transported inwards towards the centre of the nebula, and angular momentum is transported outwards towards the periphery. There are a number of different models of angular momentum transfer within accretion disks, all of which rely on shear forces of various types between adjacent layers of orbiting gas. A 'ball and spring' analogy of this process is shown in Figure 7.1. This redistribution of angular momentum within the solar nebula explains the fact that while the Sun accounts for most of the mass of the Solar System, it only carries a very small fraction of the total angular momentum.

The accretion process proceeded over tens of millions of years (around 50 million years in the case of our Sun), until the gas density near the centre was too high to allow energy to be radiated efficiently. The temperature and pressure within the newly-formed protostar then continued to increase until nuclear fusion (hydrogen burning) was initiated, and the now fully-fledged

Sun began to shine, embarking on its lifetime as a main sequence star fuelled by hydrogen burning (see Section 3.1). The internal pressure of the star at some point balanced the inward force of gravity, halting the gravitational collapse within the remainder of the nebula. In addition, radiation pressure combined with a strong *solar wind* — a continuous stream of high energy particles (mostly electrons and protons) flowing outwards from the Sun — blew most of the remaining gas out from the protoplanetary disk, effectively terminating further accretion into the growing star.

During the process of gravitational collapse and protostar formation, interesting events were also unfolding in other parts of the nebula.

7.2 The protoplanetary disk

The protoplanetary disk extended around 100 astronomical units (Earth–Sun distances) in all directions from the central proto-Sun. As we might expect, the chemical composition of the material within the disk was very similar to that of the Sun, consisting of around 98% hydrogen and helium, and around 2% heavier elements, formed via nucleosynthesis in earlier generations of stars. Ion–molecule reactions of the types described in Chapter 4 would also have ensured a healthy sprinkling of molecular species, with the most abundant being small, simple molecules such as H_2, N_2, H_2O, CO, CO_2, and CH_4. Relative abundances of the most plentiful atomic and molecular species in a dense interstellar cloud, based on data from a kinetic model by Millar and coworkers,[1] are shown in Table 7.1. Also shown are abundances for some of the major chemical components of Earth, namely iron, silicon, sulphur, and magnesium, all of which are present only in trace amounts (around one part in 10^{10}) within an interstellar cloud. Many other chemical species, which would have been present at relative abundances less than 10^{-8}, are not shown in the table.

7.3 Formation of the planets

The protoplanetary disk was subject to a large temperature gradient, with temperatures ranging from a toasty 2000 K or so near the centre of the

---

[1] T. J. Millar, J. M. C. Rawlings, A. Bennett, P. D. Brown, and S. B. Charnley, *Astron. Astrophys. Supple. Ser.*, **87**, 585–619 (1991).

Table 7.1 Relative abundances and standard melting points of atomic and molecular species in a dense interstellar cloud, as calculated by Millar and coworkers.

Species	Abundance relative to H_2	Melting point (K)
H_2	1.0	14.0
He	2.8×10^{-1}	0.95
H	1.0×10^{-4}	—
CO	1.4×10^{-4}	68
O_2	8.9×10^{-5}	54.4
O	2.5×10^{-5}	—
N_2	1.8×10^{-5}	63.2
N	7.2×10^{-6}	—
H_2O	1.9×10^{-6}	273.2
CO_2	3.6×10^{-7}	216.6
NH_3	1.1×10^{-7}	195.4
OH	9.0×10^{-8}	—
CH_4	9.0×10^{-8}	90.6
C_2H_2	7.3×10^{-8}	192
OCN	4.9×10^{-8}	—
NH_2	3.9×10^{-8}	—
NO	3.4×10^{-8}	109
CHOOH	2.8×10^{-8}	281.6
N_2O	2.3×10^{-8}	182.3
SO	1.8×10^{-8}	—
SO_2	1.6×10^{-8}	201.1
C_5	1.1×10^{-8}	—
S	5.7×10^{-10}	368.4–388.4
Mg	5.0×10^{-10}	923
Fe	3.3×10^{-10}	1811
Si	1.4×10^{-11}	1687

nebula, where the Sun was forming, to a chilly 50 K near the periphery. In the cooler reaches near the outside of the disk, the temperature was below the freezing points of most of the molecular and metallic species present, and these began to condense out of the cloud as small granules of ice and dust. As we can see from the relative abundances and melting points[2] in Table 7.1, these grains would have been made up primarily of ices containing CO, O_2, N_2, H_2O, CO_2, NH_3, and CH_4, with trace amounts of other molecular and metallic species. Closer to the centre of the protoplanetary

---

[2]Note that the melting points given in Table 7.1 have been determined at a standard pressure of 1 bar, and will therefore be slightly higher than the melting points under the low-pressure conditions in an interstellar gas cloud.

disk, where the temperatures were much higher, above the melting points and boiling points of most of the species present, only high-melting-point elements such as Fe, Mg, Si, Ca, Al, Ti, and S were able to condense out of the cloud. At the highest temperatures, over 1500 K, the most abundant solids to form would have been oxides and silicates of Ca, Al, and Ti. A little further out from the centre of the nebula, at temperatures below 1400 K, Mg and Si condensed out as Fe-free silicates, with Fe condensing as an FeNi alloy. At lower temperatures, around 750 K, FeS formed, and at still lower temperatures, below about 450 K, reactions with water led to the formation of magnetite, Fe_3O_4, and hydrated silicates. All of these compounds are abundant in the Earth.

The planets formed primarily through a process of accretion. Small microscopic grains of ice and dust drifting within the nebula collided and stuck together, eventually forming small icy or rocky bodies centimetres in diameter. These small 'space rocks' collected under the force of gravity into a thin disk with a moderately high density. The relative velocities between the particles were fairly low, and the conditions of high density and low relative velocity yielded numerous gravitational instabilities that led to relatively rapid accretion into larger bodies with dimensions of 1 km or greater, known as *planetesimals*. Asteroids and comets are thought to be 'leftover' planetesimals from this period, originating from the inner and outer reaches, respectively, of the protoplanetary disk. The large number, close proximity, and consequently chaotic orbits of these bodies resulted in many planetesimals undergoing further collisions with other similar bodies. The huge masses and correspondingly large gravitational forces at play generated collisions that released enormous amounts of energy, enough to melt and fuse the planetesimals into much larger bodies, eventually leading to the formation of the planets as they are today.

Figure 7.2 shows a spectacular image of a protoplanetary disk surrounding the young star HL Tauri, taken with the Atacama Large Millmeter Array in the Atacama Desert. This is the highest-resolution image recorded to date of a newly forming planetary system, with dark rings showing the orbits swept out by newborn planets and planetesimals as they accrete smaller bodies in their path.

The chemical compositions of the planets in our Solar System reflect the composition of the dust grains from which they were formed, and therefore the temperature in the region of the protoplanetary disk corresponding to their orbital radius. The smaller, inner planets, Mercury, Venus, Earth, and Mars, known as the *terrestrial planets*, are primarily rocky, and formed

Fig. 7.2 The protoplanetary disk surrounding the young star HL Tauri in the constellation Taurus, 450 light years from Earth. The dark rings most probably correspond to the orbits of newly forming planets as they accrete smaller bodies in their path. Credit: Atacama Large Millimeter/Submillimeter Array (ALMA) (European Southern Observatory (ESO)/National Astronomical Observatory of Japan (NAOJ)/National Radio Astronomy Observatory (NRAO)).

over a timescale of 10–100 million years. The asteroid belt is thought to have formed at the outer edge of the terrestrial region (between Mars and Jupiter) due to the influence of Jupiter's gravity on collisions between planetesimals in this region. The resulting orbital resonances led to exceptionally high-velocity collisions, in which the planetesimals shattered into smaller bodies rather than accreting into planets.

The gas giants, Jupiter and Saturn, formed just beyond the 'frost line', where the temperature becomes low enough for volatile compounds such as water, ammonia, methane, carbon dioxide, carbon monoxide to condense, and rapidly accreted a large icy core of 5–10 Earth masses. This made them sufficiently massive to accrete hydrogen and helium gas from the surrounding nebula, with accretion ceasing only when the supply of these gases had been exhausted. Along with the outer planets, Jupiter and Saturn formed much more quickly than the terrestrial planets, most probably within the first 10 million years of the evolution of the Solar System. The outer planets, Uranus and Neptune, sometimes known as 'ice giants', failed to reach a sufficient mass to accrete hydrogen and helium while there were still sufficient amounts of these gases available in the nebula. Uranus and Neptune are larger than would be predicted based on the amount of material thought to be available within the solar nebula at their current orbital radii. The most likely explanation for this is that they were initially formed in the same region of the nebula as Jupiter and Saturn, before being scattered through collisions into their present locations in the outer reaches of the Solar System. At the very outer reaches of the Solar System are found the Kuiper belt (at 30–55 astronomical units (AU) from the Sun), which contains the recently demoted dwarf planet Pluto,[3] the scattered disk (55–100 AU), and the Oort cloud (>50,000 AU). At these large distances from the Sun, accretion was too slow to lead to the formation of planets before the solar nebula dispersed, and these regions mainly contain small icy bodies. Occasionally, one of these bodies has its orbit sufficiently perturbed that it re-enters the inner Solar System and is observed from Earth as a comet.

Many of the planets have moons. These formed through a variety of processes, either accreting from gas and dust encircling the planet, forming elsewhere as a planetesimal in the early Solar System before being captured by the gravitational field of a planet, or (as in the case of Earth's moon — see Section 7.4 for further details) resulting from giant collisions between planet-sized objects.

Revisiting the list of known facts about our Solar System stated at the beginning of this chapter, we find that all of the observations are explained within the framework of the nebula model. Formation of the planets from a rotating protoplanetary disk, combined with the law of conservation of angular momentum, explains the observation that all of the planetary orbits lie in a single plane, all planets orbit the Sun in the same direction in near-circular orbits, and that this behaviour is mirrored in the orbital motion of the moons about their parent planets. It also explains the observation that the Sun and all planets apart from Venus rotate about their axes in the same direction, generally with very small tilts between their orbital and equatorial planes, as well as explaining the distribution of angular momentum between the Sun and the planets. The slow reverse spin of Venus and the large tilt of Neptune are most probably explained as resulting

---

[3] Pluto, first discovered in 1930, was for a long time considered the ninth planet. However, the discovery in 2005 of Eris, an object of a similar size to Pluto, led to a decision by the International Astronomical Union (IAU) that in order to be classified as a planet within our Solar System, an object must orbit the Sun, must be massive enough for its own gravity to make it spherical in form, and must have cleared its orbit of smaller objects. Pluto does not classify as a planet according to this definition, and was instead reclassified as a 'dwarf planet'.

from collisions with massive objects at some point during the evolution of these planets. The relatively even spacing of the planets is explained by the accretion model, and the different chemical compositions of the planets, asteroids, and comets are explained by the large temperature differential from the centre to the outside of the protoplanetary disk, and the wide range of melting points for the various chemical species present in the disk.

Our Solar System is far from unique, and many other stars possess at least one planet. Extrasolar planets will be discussed in more detail in Section 7.10.

7.4 The early Earth, and formation of the Moon

Having explored the birth of the Solar System in fairly general terms, we will now focus our attention on the Earth, and consider its evolution in more detail. The elemental composition of the Earth is shown in Table 7.2. Note that the composition of our planet is markedly different from that of the solar nebula from which it formed, in line with the temperature-dependent condensation process described in Section 7.3.

The Earth of today is very different from the Earth originally formed via the process of accretion described above. The newly-formed Earth is thought to have had only around 80% of the mass it has today, and was mostly molten due both to residual heat from the high-energy collisions between planetesimals that led to its formation and additional heat generated by radiation emitted during radioactive decay of the relatively abundant radioactive elements present (mostly ^{40}K, ^{232}Th, ^{238}U, and ^{235}U).

Table 7.2 The elemental composition of the Earth.

Element	Abundance (%)
Fe	32.1
O	30.1
Si	15.1
Mg	13.9
S	2.9
Ni	1.8
Ca	1.5
Al	1.4
Other	1.2

In the planet's molten state, gravity caused the heaviest elements (iron, nickel, and trace amounts of palladium, platinum, rhodium, and iridium) to sink to the core and the lighter elements to float to the surface, and at this point the planet consisted of a molten core surrounded by a thick liquid mantle consisting mostly of silicates. A cloud of gaseous silica surrounding the planet eventually condensed as rocks on the surface, forming the crust. The nascent Earth probably had a thin atmosphere consisting mostly of hydrogen and helium.

Early in its history, the Earth and other terrestrial planets were heavily bombarded by asteroids and comets from the outer Solar System. Many of these bodies were shifted into eccentric orbits that sampled the inner part of the Solar System through gravitational interactions with the more massive planets as they migrated and established their final orbits. The *late heavy bombardment*, as this episode in the Earth's history is known, may well have been the source of much of the carbon and water on the Earth. These species are present on Earth at much higher abundances than would be predicted simply from condensation of the solar nebula, but were abundant in planetesimals in the outer Solar System.

Around 50 million years after the Solar System began to form, the Earth is thought to have suffered one final, cataclysmic collision with another similar protoplanet around the size of Mars. The enormous amount of energy released in the collision was sufficient to fuse the metallic cores of the two protoplanets, and to eject large amounts of mantle and crust material into space, as well as blowing off the original atmosphere. The collision had a number of consequences. It increased the mass of the Earth to its current value, and also imparted a 23.5° tilt to the Earth's rotation. The ejected material entered orbit around the Earth, and rapidly accreted to form the Moon on a timescale thought to be less than 100 years. The *giant impact hypothesis* for formation of the Moon, sometimes known as the *Big Splash hypothesis*, is supported by a number of observations, perhaps the most persuasive being the very similar chemical compositions of the Earth and the Moon. The protoplanet that gave birth to our Moon has been christened Theia, after the Greek Titan goddess, mother of the Moon goddess Selene.

As well as the collision with Theia leading to the formation of the Moon, the axial tilt imparted to the Earth's rotation axis relative to the ecliptic (the plane of the Earth's orbit around the Sun) is responsible for the Earth's seasons. As illustrated in Figure 7.3, the tilt has the consequence that the amount of sunlight falling on the Northern and Southern hemispheres

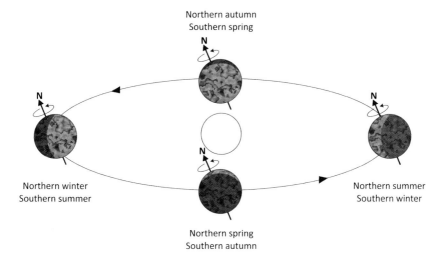

Fig. 7.3 The Earth's 23.5° tilt gives rise to the seasons with the Northern and Southern hemispheres being exposed to differing amounts of sunlight as the Earth orbits the Sun.

of the Earth varies significantly at different points in the Earth's orbit around the Sun. The section of the orbit in which the Northern hemisphere receives considerably more sunlight than the Southern hemisphere corresponds to the Northern summer and Southern winter. The sections in which both hemispheres receive similar amounts of sunlight correspond to Northern spring/Southern autumn or vice versa, and the section in which the Southern hemisphere receives more sunlight than the Northern hemisphere corresponds to Southern summer/Northern winter.

7.4.1 The Moon's orbit and tidal locking

The orbit of the Moon has changed considerably since the Moon was first formed, and is still evolving. The Moon was originally much closer to the Earth, and the orbital radius is still increasing by around 3–4 cm per year. This is primarily a result of *tidal forces*. These same forces are also responsible for the *tidal locking* between the Moon and the Earth that causes the Moon's rotational period to match its orbital period, such that one face of the Moon permanently faces the Earth. While the details of the Moon's orbital motion are quite complicated, the motion can be explained relatively easily to first-order using a simple model that considers the gravitational interaction between the Earth and the Moon in the context of conservation

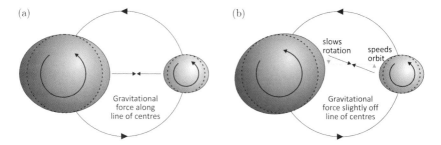

Fig. 7.4 (a) Tidal bulges are caused by the gravitational attraction between a planet and its moon. In a tidally locked system, both planet and moon rotate at the same frequency as they orbit, and the tidal forces between the bulges act along their line of centres. (b) If the planet rotates faster than it orbits (currently the case for the Earth–Moon system), then while gravity acts primarily along the planet–moon axis, the 'additional' tidal forces between the bulges are off axis. This causes the planet's rotation to slow and the moon's orbital velocity (and therefore orbital radius) to increase, until eventually the rotational and orbital periods of both planet and moon are matched. Our Moon is already tidally locked to the Earth, but it would take 50 billion years for the Earth to become tidally locked to the Moon. If rotation is slower than the orbital motion, tidal forces act in reverse to speed up the rotation and slow the orbital velocity.

of angular momentum. The gravitational attraction between the Earth and the Moon leads to a distortion in the shape of both bodies, causing them to become slightly elongated in the direction corresponding to the Earth–Moon axis, and slightly compressed in the direction perpendicular to this axis (see Figure 7.4). The distortions are known as *tidal bulges*, and it is the gravitational forces on these bulges that lead both to tidal locking and to the gradual increase in the Earth–Moon orbital radius with time.

If the Earth and Moon were stationary, the gravitational attraction between them would always act along the same axis and the tidal bulges would be static. However, in reality, the bulges must respond to changes in the direction of the gravitational force as the Earth and the Moon orbit about each other and rotate about their individual axes. As shown in Figure 7.4, if the Moon rotates faster than it orbits, which was in fact once the case, the tidal bulges are always slightly ahead of the Earth–Moon axis in the direction of rotation, and gravitational forces on the bulges act to slow down the speed of rotation until the rotational and orbital periods match. In fact, the force on the 'near-side' bulge acts to slow the speed of rotation, and the force on the 'far-side' bulge acts to speed up rotation, but the $1/r^2$ dependence of the gravitational force on distance means that the force on the 'near-side' bulge is slightly stronger, leading to a net slowing down of the speed of rotation. Orbiting bodies whose rotational period is initially longer than their orbital period can also become tidally locked. In this case, the tidal bulges lag behind the planet–moon axis and the tidal forces act to speed up the rotation until the rotational and orbital periods match.

Tidal forces also act on the Earth's tidal bulges, and are causing the Earth's rotation to slow down gradually over time. As a result, each Earth year is around 15 μs longer than the one before. The Earth would become tidally locked to the Moon in around 50 billion years, but this is highly unlikely to happen given that the Sun is expected to reach the end of its lifetime as a main sequence star in around 5 billion years, at which point it will evolve into a red giant and engulf both the Earth and the Moon. Pluto and its moon Charon, which is similar in size to Pluto and orbits the planet relatively closely, provide an example of a planet–moon system in which both bodies are tidally locked to each other.

A further consequence of tidal interactions results from conservation of angular momentum. Since the total angular momentum of the Earth–Moon system must be conserved, any decrease in the rotational angular momentum of the Earth or the Moon must be accompanied by an increase in the angular momentum associated with their orbital motion. The magnitude of the orbital angular momentum (see Section 5.4.9) is given by $\mu v r$, where μ is the reduced mass of the Earth and the Moon, v is their relative velocity (approximately equal to the orbital speed of the Moon in this case), and r is their separation. Transfer of angular momentum from the Earth's spin to the Earth–Moon orbital angular momentum results in an increase in the orbital radius. As noted at the beginning of this section, the Earth–Moon distance is currently increasing by around 3–4 cm per year.

7.5 The layered structure of the Earth

Following the giant impact that led to the formation of the Moon, and other smaller impacts, the Earth gradually cooled again over a timescale of around 100 million years into its present layered structure, shown schematically in Figure 7.5. The structure has been determined by studying the speed at which seismic waves from earthquakes travel through the Earth. These waves undergo refraction and reflection at the various interfaces within the Earth's structure, allowing the measured time of travel of the waves to be used to infer features of the structure. To the best of our current knowledge, the structure consists of a hot but (probably) solid inner core,

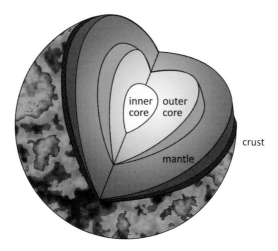

Fig. 7.5   The layered structure of the Earth.

a liquid outer core, a viscous mantle, and a solid crust, surrounded by a gaseous atmosphere. We now consider the various layers in turn.

7.5.1   The core and the Earth's magnetic field

The Earth's solid inner core has a radius of around 1220 km and is composed primarily of iron ($\sim$80%), with some nickel and trace amounts of other elements. The chemical composition of the outer core, which extends to around 3400 km from the Earth's centre, is thought to be similar, with the somewhat lower pressures in this region causing it to behave as a liquid rather than a solid.

The metallic core is responsible for generating the Earth's magnetic field. While it is tempting to propose that magnetisation of the solid core occurred at some point in the early evolution of the Earth and is the source of the magnetic field, the temperature in the core is well above the Curie temperature of iron, making permanent magnetisation impossible. In addition, a solid core cannot easily explain the fact that the Earth's magnetic poles are not static: their positions are currently drifting at a speed of several tens of kilometres per year, and the field is also known to reverse its polarity completely from time to time, at intervals of between a few thousand and a few million years (around 200,000 on average). Taken together, these observations imply a dynamical, rather than a static, source for the field. While the Earth's magnetic field is still not completely understood, the best current explanations invoke a 'geodynamo', in which the magnetic field is generated by electrical currents within the moving conductive fluid that makes up the outer core. The temperature gradient across the liquid outer core sets up convection currents in the liquid iron, which couple with the rotational motion of the Earth to yield helical flow patterns about the Earth's rotation axis. The flow dynamics and resulting magnetohydrodynamic interactions are extremely complex, and the situation is complicated still further when one considers the effects of the solid crystalline iron core. However, computer simulations of the dynamics[4] have been able to reproduce both the magnitude and shape of the Earth's essentially dipolar magnetic field, as well as its small tilt of around 11° relative to the Earth's rotation axis. The simulations have even predicted polarity reversals on approximately the same timescales as observed, lending considerable support to the geodynamo description.

The Earth's magnetic field (known as the *magnetosphere*) protects the atmosphere and surface of the planet from the stream of high energy charged particles that make up the Solar wind, and has therefore been vitally important in allowing complex chemistry to develop on our planet. As shown in Figure 7.6, the magnetic dipole field of the Earth deflects energetic charged particles from the Sun into curved trajectories around the planet, preventing them from penetrating deep into the atmosphere or reaching the surface. At the magnetic poles, electronic excitation and ionisation of atmospheric oxygen and nitrogen by cosmic ray particles accelerated into the upper atmosphere by the Earth's magnetic field often generates intense fluorescence, visible from Earth's surface as the *Aurora Borealis* (Northern lights) and *Aurora Australis* (Southern lights).

7.5.2 The mantle

The Earth's mantle is almost 3500 km thick, and accounts for around 84% of the planet's volume. It is composed mainly of a coarse-grained igneous[5]

---

[4]G. A. Glatzmaier and P. H. Roberts, A three-dimensional convective dynamo solution with rotating and finitely conducting inner core and mantle, *Phys. Earth Planet. Inter.*, **91**, 63–75 (1995); G.A. Glatzmaier and P.H. Roberts, A three-dimensional self-consistent computer simulation of a geomagnetic field reversal, *Nature*, **377**, 203–209 (1995).

[5]Igneous rocks are formed through the cooling and solidification of magma or lava. Most are formed below the Earth's crust, but they can also be formed at the crust in lava flows. There are two other classes of rock. Sedimentary rocks are formed either through precipitation from the oceans or through accumulation and solidification of rock

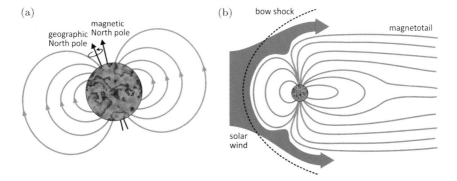

Fig. 7.6 (a) Convection currents in the Earth's core give rise to a dipolar magnetic field, currently aligned at a small angle to the Earth's rotation axis; (b) the Earth's magnetic field deflects the solar wind, protecting the Earth's atmosphere and surface. The solar wind is deflected at an imaginary surface known as the bow shock, and the interaction between the solar wind and the magnetic field alters the shape of the dipolar field as shown.

rock called peridotite (see Figure 7.7(a)), which consists mainly of silicates and the elements magnesium and iron. These are present in the rock mostly in the form of olivine and pyroxene minerals. Olivines (see Figure 7.7(b)), which make up the major component at depths down to around 400 km, have the general chemical formula $(Mg^{2+}, Fe^{2+})_2SiO_4$, and may be known to some as the gemstone peridot. The ratio of Mg^{2+} to Fe^{2+} ions within the olivine 'solid solution' structure is variable, with compositions ranging from Mg_2SiO_4 (known as forsterite) to Fe_2SiO_4 (fayalite). Deeper into the Earth's mantle, below depths of around 400 km, the enormous pressures cause the olivine to adopt different ('polymorphic') crystal structures, known as wadsleyite and ringwoodite. Only the Mg_2SiO_4 version of wadsleyite is thermodynamically stable under these conditions. Pyroxenes are chemically more complicated than olivines. They have the general formula $XY(Si,Al)_2O_6$, where X can be Ca^{2+}, Na^+, Fe^{2+}, Mg^{2+}, Zn^{2+}, Mn^{2+}, or Li^+, and Y can be any one of a number of (usually) smaller ions, e.g. Cr^{3+}, Al^{3+}, Fe^{3+}, Mg^{2+}, Mn^{2+}, Sc^{3+}, Ti^{3+}, Va^{3+}, and occasionally Fe^{2+}. The chemical composition of the mantle is very similar to that of the Moon, lending weight to the 'giant impact hypothesis' described in Section 7.4.

---

fragments, minerals, and organisms at the Earth's surface. Metamorphic rocks are formed when an igneous or sedimentary rock is subjected to high temperature and pressure, sufficient to alter its chemical structure.

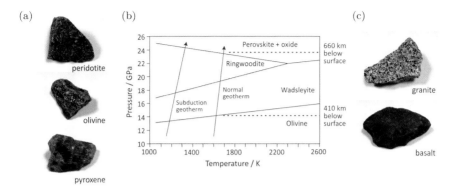

Fig. 7.7 (a) Mantle minerals: Peridotite consists primarily of olivine and pyroxene minerals; (b) phase diagram illustrating the phase transitions from olivine to wadsleyite and ringwoodite that occur deep in the Earth's mantle (adapted from G. R. Helffrich and B. J. Wood, The Earth's Mantle, Nature, **412**, 501–507 (2001)); (c) crust minerals: the Earth's crust consists primarily of granite and basalt rock.

Contrary to most popular understanding, the Earth's mantle is solid rock, not liquid. However, the high-temperatures in this region of the Earth, ranging from a few hundred degrees celcius near the crust to over 4000 degrees at the mantle–core boundary, make the silicate material sufficiently ductile that it can flow on geological timescales. The very high pressures deep in the Earth result in the lower part of the mantle flowing less easily than the upper part. To give some idea of the timescales involved, we can consider the viscosity of the mantle in comparison with a number of familiar liquids. Water at 20°C has a viscosity of 1.002×10^{-3} Pa s, molasses has a viscosity of around 5–10 Pa s, and molten glass has a viscosity of 10–1000 Pa s. In comparison, the viscosity of the mantle varies from 10^{21} to 10^{24} Pa s, depending on depth. The mantle is often subdivided into two-layers based on the flow properties of the mantle rock. The top part of the mantle, above about 250 km, is known as the *asthenosphere*. Though made of solid rock, this region of the mantle is relatively soft and flows easily. The deeper part of the mantle is known as the *mesosphere*. The rock within this region is still capable of flowing, but much less easily than in the asthenosphere.

Despite its slow rate, flow within the mantle is very important in shaping the surface of the Earth. The temperature differential across the mantle leads to (extremely slow) convection currents that cause hot material to rise through the mantle and colder material to sink over a timescale of tens of millions of years. This convective motion drives the motion of tectonic plates, and is therefore the ultimate origin of earthquakes, mountain ranges, and volcanos. Plate tectonics and their consequences will be considered in more detail in Section 7.5.3. Magma, or molten mantle rock, often forms in rising mantle plumes under conditions of high temperature and low pressure within a few kilometres of the Earth's surface. Depending on the integrity of the crust directly above the resulting *magma chamber*, the magma may either erupt from the chamber through a fissure in the crust, flowing as volcanic lava, or, if the overlaying crust remains intact, it may simply cool and recrystallise within the mantle.

7.5.3 The crust

The crust makes up less than 1% of the Earth's volume. The Earth's crust is far from uniform, exhibiting a variety of topological features including ocean basins, continents, mountain ranges, and volcanos. There are known to be two distinct types of crust: continental crust, covering around 40% of the planet, consists primarily of silicates and alumina, and is typically 30–50 km thick; while oceanic crust has a composition much more similar to the mantle, and is typically only 5–10 km thick. Continental crust is also known to be much older than oceanic crust, with the oldest rocks dated at over 4 billion years of age. Oceanic crust, in contrast, is typically regenerated with a life cycle of around 180 million years. Explaining these observations requires us to understand a little about tectonic plate motion.

Convection currents within the mantle (see Section 7.5.2) have had the consequence that rather than forming as a single solid shell, the Earth's crust is in fact made up of a number of separate pieces known as tectonic plates, which are able to move relative to each other. Tectonic plates consist of sections of crust together with the uppermost part of the mantle, which together make up the lithosphere. The lithosphere 'floats' on the underlying asthenosphere, with individual tectonic plates moving as a result of convection currents within the mantle. The lithosphere is relatively brittle, and is prone to fracture when subjected to large forces.

Plate tectonic motion is thought to have begun between 3 and 4 billion years ago.[6] The formation of ocean basins, continents, volcanos, and mountain ranges on the Earth's surface can all be understood in terms of the motion of tectonic plates, and in particular by considering the processes occurring at *divergent plate boundaries*, where tectonic plates move

---

[6]See for example D. Bercovici, and Y. Ricard, *Nature*, **508**, 513–516 (2014).

away from each other, and *convergent plate boundaries*, where they move towards each other. A final type of boundary is known as a *transform fault boundary* (the San Andreas fault provides an example of such a boundary), and occurs where two tectonic plates slide past each other. All three types of boundary generate powerful earthquakes, with divergent and convergent boundaries often also resulting in volcanism.

7.5.3.1 Divergent plate boundaries

Divergent plate boundaries are responsible for the formation of ocean basins, and for the continual recycling of oceanic crust. When convection currents within the mantle cause two plates to move apart from each other, the pressure on the underlying mantle is reduced, prompting a phase transition in which the peridotite melts into magma. The magma erupts from the asthenosphere as basaltic lava at an *oceanic ridge* or *spreading centre* , creating new ocean floor as it spreads away on either side of the plate boundary, and subsequently cools and solidifies. The erupting magma tends to comprise the less dense components of the mantle, mainly silicates and alumina, along with any elements that do not incorporate readily into mantle minerals (these are known as *incompatible elements*). The heavier residue left behind in the mantle is enriched in iron and magnesium. Many of the incompatible elements eventually make their way into the continents at convergent plate boundaries.

While divergent plate boundaries usually occur within ocean basins, it is also possible for continental crust to fracture and split apart. The formation of a *rift zone* in this way leads to the creation of new ocean basins.

7.5.3.2 Convergent plate boundaries

When two plates move towards each other, one plate must sink below the other in a process known as *subduction*. All convergent boundaries generate powerful and frequent earthquakes (for example, the 'Ring of Fire' in the Pacific Ocean basin is the result of subduction at a number of convergent plate boundaries), but the topological consequences depend on whether the converging plates consist of two sections of oceanic crust, two sections of continental crust, or one section of each type of crust.

When two sections of oceanic crust meet at a convergent plate boundary, the subducting (sinking) plate is pushed down into the mantle to depths at which melting occurs. The magma rises to the surface and erupts to form 'island arcs'. Examples include the Philippines, the Solomon Islands, and the Japanese Archipelago.

When an oceanic plate meets a continental plate, the higher density of the oceanic plate will cause it to be subducted beneath the continental plate. Similar to the process occurring when two oceanic plates meet, the subducted oceanic plate melts, and magma rises to the surface, this time forming a chain of volcanos on the continental margin known as a *volcanic arc*. Examples include the Andes mountain range in South America, the Cascade Volcanoes in North America, and the Aleutian range in Southwest Alaska.

Finally, convergence of two continental plates tends to lead to the formation of high mountain ranges, with the Himalayas providing a spectacular example.

Armed with a basic understanding of plate tectonic motion and subduction, most of the surface features of the Earth, as well as the differing mineral compositions of rocks within the continental and oceanic crust, are relatively easily explained. Continental crust consists primarily of granite rocks, whereas oceanic crust is primarily composed of basalt. Both rock types are a mixture of oxides, but their chemical composition differs considerably, as shown in Table 7.3. The much greater age of rocks within the continental crust is explained by the fact that the lower density of continental crust makes it too buoyant to undergo subduction. For the same reason, the volume of continental crust has grown slowly, at the expense of oceanic crust, over a timescale of several billion years. The incompatible elements

Table 7.3 Approximate average composition of granite and basalt rocks in the Earth's crust.

Component	Granite %	Basalt %
SiO_2	72.04	49.97
Al_2O_3	14.42	15.99
K_2O	4.12	1.12
Na_2O	3.69	2.96
CaO	1.82	9.62
FeO	1.58	7.24
Fe_2O_3	1.22	3.85
MgO	0.71	6.84
TiO_2	0.30	1.87
P_2O_5	0.12	0.35
MnO	0.05	0.20

Source: Data from M. G. Best, *Igneous and Metamorphic Petrology*, 2nd edn. (Wiley-Blackwell, 2002).

from the mantle that make their way into the continental crust also play an important role in the evolution of the Earth's surface. These include radioactive potassium, thorium, and uranium, which heat and soften the crust, allowing it to spread sideways.

While most transfer of matter at the Earth's surface is from oceanic crust to continental crust as a result of plate tectonics, other processes are also important. Weathering, erosion, and sedimentation, considered in more detail in Section 7.7, can cause material to be transferred in the opposite direction, from the continents to the ocean floor. However, even sand and clay washed from the continents into the sea will eventually find its way back to the continents via tectonic motion.

7.5.4 The primordial atmosphere

In Section 7.4, we noted that in its early stages of formation, the Earth probably had a thin atmosphere consisting primarily of hydrogen and helium. This was either blown off in the cataclysmic collision with a planetesimal that led to formation of the Moon, or simply drifted away. Earth's gravity is not strong enough to hold these lighter gases, and before the formation of the Earth's core and the subsequent generation of the Earth's magnetic field, strong solar winds would also have made it difficult for an atmosphere to be retained.

As the Earth cooled into its present layered structure, a second atmosphere formed through volcanic outgassing. The composition of this primordial atmosphere was probably similar to the present-day atmosphere of Venus, a similarly-sized rocky planet whose atmosphere also formed through outgassing. Venus's atmosphere is dominated by carbon dioxide (around 96.5%) with a small amount of nitrogen (3.5%). Other gases, present at concentrations of tens of parts per million or less, include sulphur dioxide, argon, water vapour, carbon monoxide, helium, neon, hydrogen chloride, and hydrogen fluoride. The early atmosphere was very different from today's atmosphere, which consists of around 80% N_2 and 20% O_2, for reasons we shall explore in Section 7.8.

7.6 Oceans and tides

There is still considerable controversy over the origin of water on Earth, but it is fairly clear from the geologic record that large quantities of surface water were present from around 3.8 billion years ago. During the very early stages of the Earth's formation, the temperature was too high for volatile molecules such as water, carbon dioxide, and methane to condense. However, a number of plausible mechanisms have been proposed in which large amounts of water were present on Earth from the very early stages of its formation.[7] Such models invoke a massive hydrogen-rich atmosphere that reacted with oxides in the Earth's molten surface to produce large quantities of water, with the high pressure of the hydrogen atmosphere preventing the water from boiling away at the high surface temperatures of this period. Recent discoveries[8] of large reservoirs of water deep within the Earth's mantle also offer the intriguing possibility that water was in fact present from the very early stages of our planet's formation, chemically bound to minerals, and has simply outgassed from the planet over geological timescales to form the oceans.

The alternative, and generally more widely accepted hypothesis, is that water and carbon-containing compounds arrived on Earth during the late heavy bombardment. During this period, the Earth was bombarded by a large number of comets and meteorites from the outer Solar System, where water and carbon compounds did condense. Though icy comets might appear to be the prime candidates for delivering water to Earth, representing huge reservoirs that could easily deposit sufficient quantities of water to form the Earth's oceans, measurements of the isotopic ratios of D_2O to H_2O in water from Earth and from several comets appear to indicate that comets were at least not the sole source of Earth's water. Very recent measurements from the Rosetta orbiter,[9] for example, indicate that the comet 67P Churyumov–Gerasimenko has a $D_2O:H_2O$ ratio three times higher than that found in terrestrial water. The terrestrial isotopic ratio is much better matched by that of water from rocky carbonaceous chondrites, a type of asteroid, making these perhaps the most likely source of terrestrial water.

The majority of Earth's water fills the ocean basins (see Section 7.5.3 for a description of the surface topography of the Earth). However, this water is

---

[7] H. Genda and M. Ikoma, Origin of the Ocean on the Earth: Early evolution of water D/H in a hydrogen-rich atmosphere, *Icarus*, **194**, 42–52 (2008).

[8] B. Schmandt, S. D. Jacobsen, T. W. Becker, Z. Liu, and K. G. Dueker, Dehydration melting at the top of the lower mantle, *Science*, **344**(6189), 1265–1268 (2014).

[9] K. Altwegg *et al.*, 67P/Churyomov–Gerasimenko, a Jupiter family comet with a high D/H ratio, *Science*, **347**(6220), 387 (2015).

not static. Firstly, gravitational forces exerted by the Moon and Sun result in a twice-daily cycle of high and low tides. Tidal forces were considered in Section 7.4.1. In this section, we considered distortion of the planet itself by the gravitational force between the Earth and the Moon. Water is of course much more easily distortable than rock, leading to marked tidal bulges at either pole of the Earth–Moon axis. The Sun's gravity also exerts a force on the Earth's water, and the amplitude of the tides changes as the Moon orbits the Earth and the Earth orbits the Sun according to the relative alignment of the Earth–Moon and Earth–Sun axes. The highest (spring) tides occur when the axes are aligned with each other and the gravitational pull of the Sun and Moon reinforce each other. Conversely, the lowest (neap) tides occur when the two axes are at right angles and the Sun and Moon are 'pulling' in perpendicular directions.

The second mechanism for redistribution of sea water is of course the hydrologic cycle, which is responsible for removing water from the oceans through evaporation to form clouds, and depositing it in lakes and rivers either directly through rainfall into these reservoirs, or indirectly through rainfall on land and subsequent run-off. While sea water has high concentrations of salts dissolved from the surface rock, these are mostly left behind when sea water evaporates to form clouds. The result of this 'natural distillation process' is that lakes and rivers contain water with much lower salt concentrations than the oceans.

Interestingly, when the Earth's oceans first formed, the Sun was around 30% dimmer than it is now, and was therefore much cooler. We might expect that under such conditions the Earth's oceans should have frozen. However, this was clearly not the case. The explanation lies in the composition of the atmosphere at the time, which was rich in the 'greenhouse gases' CO_2, water, and methane. To understand the Greenhouse effect, we need to consider the balance between the incoming flux of Solar radiation at the surface of the Earth and the outgoing flux irradiated into space. This balance is affected quite considerably by the Earth's atmosphere. The temperature of the surface of the Sun is around 5800 K, and the incoming Solar radiation spectrum at the top of the Earth's atmosphere is well approximated by the corresponding black-body[10] spectrum, peaking in the visible region

---

[10] A black body is a perfect absorber and emitter of radiation. The emission spectrum, $I(\nu,T)$, of a black body depends only on its temperature, T, and is given by Planck's law, $I(\nu,T) = (2h\nu^3/c^2)(\exp(h\nu/k_B T) - 1)^{-1}$, where ν is the frequency of the emitted radiation, h is Planck's constant, c is the speed of light, and k_B is the Boltzmann constant.

and tailing off into the ultraviolet and infrared regions of the electromagnetic spectrum. Only around 70% of the radiation arriving at the top of the atmosphere reaches the Earth's surface, with the major loss processes being reflection by clouds and absorption by atmospheric gases, mostly in the UV and IR regions of the spectrum. The Solar radiation reaching the Earth's surface warms the planet, and it is relatively straightforward to predict the resulting surface temperature by assuming that the Earth will act as a black body at this temperature, and balancing the incoming and outgoing energy. This treatment predicts a mean surface temperature for the Earth of around 255 K, or $-18°C$, which is clearly much lower than the actual mean surface temperature of around 288 K, or $15°C$. The reason for this discrepancy lies in the fact that a black body in this temperature range emits radiation primarily in the infrared region of the spectrum, where many relatively abundant atmospheric gases — particularly CO_2, water, and methane — absorb strongly. Instead of being radiated into space, much of the energy radiated from the Earth's surface becomes trapped in the lower 10 km or so of the Earth's atmosphere, insulating the surface and warming it to a greater degree than predicted by the simple energy balance model above. In the primordial Earth, the strong Greenhouse Effect resulting from the high levels of the CO_2, water, and methane in the atmosphere allowed liquid water to exist during this critical phase of the planet's development, despite the considerably lower incoming Solar flux from the young Sun.

7.7 Erosion and weathering

The presence of the oceans and an atmosphere, and subsequently wind and rain, facilitated further physical and chemical mechanisms for modifying the surface of the Earth. Rain, wind, and flowing water drive a variety of *erosion* and *sedimentation* processes, which transport soil and rock from one location on the Earth's surface to another. Erosion processes occurring over geological timescales have shaped river beds, coastlines, rock formations, mountain ranges, scree slopes, deserts, and many other features of the Earth's surface, as well as erasing the evidence of all but a hundred or so of the many thousands of meteor impacts suffered by the Earth over its lifetime.

An atmosphere rich in CO_2 surrounding a planet possessing large amounts of surface water also led to a further mechanism for altering the chemistry of Earth's surface, namely chemical weathering. Carbon dioxide dissolves readily in surface water and cloud droplets to form carbonic acid, a weak acid that partly dissociates in solution to form hydrogen ions, H^+, and bicarbonate ions, HCO_3^-.

$$CO_2 + H_2O \rightarrow H_2CO_3(aq) \tag{7.1}$$

$$H_2CO_3(aq) \rightleftharpoons H^+(aq) + HCO^-(aq) \tag{7.2}$$

As a result of this process, rainwater typically has a pH of around 5.5. While not strongly acidic, over geological timescales this slightly acid 'reagent' can cause considerable chemical change at the surfaces with which it comes into contact. The dissolved hydrogen ions react readily with most common silicate and carbonate minerals, predominantly yielding clays and various soluble ions (e.g. Na^+, K^+, Ca^{2+}, Fe^{2+}). Most clays, formed from sheets of aluminosilicates, result from weathering of feldspar minerals. For example:

$$4\,KAlSi_3O_8 + 4H^+ + 2H_2O \rightarrow 4\,K^+ + Al_4Si_4O_{10}(OH)_8 + 8SiO_2$$
$$\text{(orthoclase feldspar} + 4H^+ + 2H_2O \rightarrow 4\,K^+ + \text{kaolinite} + \text{quartz)} \tag{7.3}$$

Another common weathering reaction is that of wollastonite (calcium silicate, $CaSiO_3$):

$$CaSiO_3 + 2H^+ + H_2O \rightarrow Ca^{2+} + H_4SiO_4 \text{ (silicic acid)} \tag{7.4}$$

In the oceans, dissolved calcium ions and bicarbonate are precipitated out as calcium carbonate, and silica from the silicic acid is precipitated out as silicon dioxide. On the present-day Earth, these reprecipitation processes are carried out primarily by marine organisms: calcium carbonate is used by a wide variety of marine species, including corals and shellfish, while silica is used by planktonic diatoms. Eventually the calcium carbonate finds its way to the sea floor, where it forms layers of limestone.

Weathering processes of the type described above are responsible for removing around one gigatonne of CO_2 per year from the atmosphere, providing a sizeable CO_2 sink. However, as we shall see in Section 7.8, they do not provide the explanation for the transformation of the primordial atmosphere described in Section 7.5.4 into the present-day atmosphere.

In addition to reshaping the surface of our planet through the processes of weathering and erosion, the presence of liquid water, an ideal solvent for complex solution-phase chemistry, set planet Earth on a chemical trajectory unlike that of any other known planet.

7.8 Life and the oxygen atmosphere

We have seen already that, given sufficient time, complex organic molecules form readily even in the inhospitable low-temperature and low-pressure-environment of an interstellar gas cloud (see Chapter 4). It should therefore come as no surprise that in the relatively warm waters of Earth, complex organic chemistry developed on a rapid timescale. A famous example of complex organic chemistry occurring under conditions similar to those on early Earth is provided by Urey and Miller's famous 'primordial soup' experiments.[11] Performed in 1952, these experiments demonstrated that many of the molecules that constitute the building blocks of life, including a large number of different amino acids, could be generated readily from a mixture of water, hydrogen, methane, and ammonia, given an appropriate energy source, in their case an electric discharge. In a second example of complex chemistry arising from an innocuous starting point, relatively simple molecules consisting of a hydrophilic 'head group' and a hydrophobic 'tail' readily self-assemble in aqueous solution to form a variety of single-layered and double-layered spherical micelles. Some examples are shown in Figure 7.8. A double-layered spherical micelle is essentially a prototype cell membrane, illustrating that what may at first appear to be extremely complex chemical structures can in fact form spontaneously from very simple building blocks, so long as they are present at sufficiently high concentrations.

There are many different theories as to the detailed origins of life,[12] and a full examination of this fascinating topic is well beyond the scope of this book. However, it is clear that relatively early on in our planet's history, the chemical complexity on Earth had increased to the point at which molecules capable of replicating themselves had developed. Once this threshold had been crossed, replication provided a mechanism by which molecules could evolve towards ever greater complexity. The first single-celled organisms,

---

[11] S. L. Miller, Production of amino acids under possible primitive Earth conditions, *Science*, **117**, 528–529 (1953).
[12] For a fairly recent and accessible overview, see for example P. Davies *The Origin of Life* (Penguin, 2003).

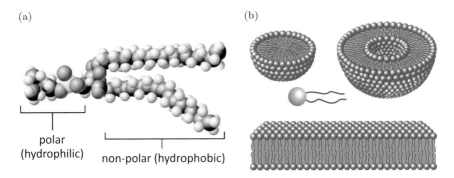

Fig. 7.8 (a) A phospholipid consists of a polar, hydrophilic 'head group' and a non-polar, hydrophobic 'tail group'; (b) these relatively simple building blocks can self-assemble into a variety of complex structures in solution, depending on the shape and size of their head and tail groups. Structures adopted include single-layered and double-layered spherical micelles and lipid bilayers.

known as cyanobacteria or (somewhat inaccurately) 'blue-green algae', emerged between 2.1 and 2.7 billion years ago, and still flourish today. Cyanobacteria are very simple single-celled prokaryotic organisms, consisting of a cell containing a single strand of DNA, but no nucleus. Cyanobacteria live in the upper reaches of the oceans, and capture energy from the Sun via photosynthesis. While the detailed mechanism of photosynthesis is extremely complex, involving a series of protein-mediated electron transfer reactions,[13] the net result is to convert carbon dioxide and water into carbohydrate and oxygen.

$$2n\text{CO}_2 + 2n\text{H}_2\text{O} + h\nu \rightarrow 2(\text{CH}_2\text{O})_n + 2n\text{O}_2 \qquad (7.5)$$

Initially, the O_2 generated through photosynthesis reacted rapidly with vast stores of iron dissolved in the oceans and present at the surface, and was eventually subsumed into the Earth's crust and mantle. However, as the number of cyanobacteria increased, eventually the rate of O_2 production exceeded its rate of consumption, and oxygen began to build up in the atmosphere. The buildup was not steady with time, and there were several steps as various reservoirs for O_2 became saturated. The evolving atmosphere triggered at least one ice age, or 'snowball Earth' event. There is considerable debate as to the severity of this event and the underlying

---

[13] R. E. Blankenship, *Molecular Mechanisms of Photosynthesis* (Wiley-Blackwell, 2001).

mechanism or mechanisms responsible,[14] but the spectacular cooling of the Earth during this period may at least be plausibly explained by considering the effect of the evolving atmosphere on incoming radiation from the Sun. High O_2 levels would have oxidised methane to CO_2, a weaker greenhouse gas, reducing the ability of the atmosphere to absorb the infrared radiation emitted from the Earth's surface, and allowing a greater fraction to radiate into space. Once the Earth's surface had frozen and was covered with snow and ice (the temperature at the Equator during the snowball Earth episode is thought to have been similar to that of the Antarctic today), the effect was compounded by the resulting much higher reflectivity (albedo) of the surface to sunlight. Under these conditions, a much higher proportion of sunlight was reflected from the surface and much less was absorbed than in the case of a warmer Earth.

The resulting cooling of the atmosphere led to a reduction both in the rate of oxidation in the atmosphere and the in rate of chemical weathering at the surface, increasing the atmospheric concentrations of both methane and CO_2. Methane and CO_2 emitted from volcanoes boosted these concentrations further, until eventually the atmosphere trapped sufficient reflected radiation and warmed sufficiently to melt the surface ice and thaw the surface of the Earth. It is clear even from this short discussion that many complex interlinked factors were at play during the evolution of the modern atmosphere, but eventually a stable equilibrium atmospheric composition was achieved. Modern levels of O_2 in the atmosphere were reached around 600 million years ago.[15]

The production of an oxygen atmosphere had profound consequences for the development of further life on Earth. In the upper reaches of the Earth's atmosphere, reactions involving atomic and molecular oxygen formed a layer of ozone, O_3. The ozone layer shielded the surface of the Earth from UV wavelengths that would be extremely damaging to life, and made the Earth's surface habitable.

While the detailed chemistry and physics of the ozone layer is highly complex, and beyond the scope of this text,[16] the basic chemical reactions that led to formation of a layer of ozone in the stratosphere were elucidated

---

[14] See for example P. F. Hoffman and D. P. Schrag, The snowball Earth hypothesis: Testing the limits of global change, *Terra Nova*, **14**(3), 129–255 (2002).

[15] R. P. Wayne, *Chemistry of Atmospheres*, 3rd edn. (Oxford University Press, 2000).

[16] The detailed chemistry of the ozone layer is covered in the second text in this series: *Atmospheric Chemistry: From the Surface to the Stratosphere*.

by Sydney Chapman in 1930, and constitute the *Chapman cycle*:

$$O_2 + h\nu \rightarrow O + O$$
$$O + O_2 + M \rightarrow O_3 + M$$
$$O_3 + h\nu \rightarrow O + O_2$$
$$O_3 + O \rightarrow O_2 + O_2 \tag{7.6}$$

The first two steps are key: efficient formation of ozone requires both a high flux of photons with wavelengths shorter than 242 nm, the energetic threshold for photolysis of the O=O bond, and a high abundance of O_2. At high altitudes the abundance of short-wavelength photons is high, but the abundance of O_2 is low as a result of the exponential decay in atmospheric pressure away from the Earth's surface. At low altitudes the abundance of O_2 is high, but most of the short-wavelength photons have already been absorbed by O_2 at higher altitudes. There is therefore a relatively narrow altitude range, lying between around 20 and 30 km from the Earth's surface, within which the rates of the reactions leading to ozone formation reach a maximum, and a layer of ozone is formed. The ozone concentration within the ozone layer is around 2–8 parts per million (ppm), compared with around 0.3 ppm at the Earth's surface.

Ozone is extremely efficient at absorbing UV radiation. If all of the ozone in the ozone layer was brought down to sea level, it would form a layer only 3 mm thick on the Earth's surface at atmospheric pressure. However, even this small amount of ozone absorbs virtually all of the incoming solar radiation in the wavelength range from 200 to 310 nm, with shorter wavelengths being absorbed effectively by O_2. Such wavelengths are extremely damaging to biological organisms, and the formation of the ozone layer is one of the key developments that made life on the Earth's surface possible.

In addition to generating the ozone layer, the presence of high concentrations of atmospheric O_2 led to the evolution of aerobic respiration. While this may have occurred first in prokaryotic organisms such as cyanobacteria, which hitherto had generated their energy via either photosynthesis or anaerobic respiration, the fossil record shows that the development of aerobic respiration coincided with the rise of eukaryotic organisms. The cells of a eukaryote contain DNA surrounded by a membrane–enclosed nucleus. While prokaryotic organisms have never evolved beyond the microbial or single-celled stage, many eukaryotes became multicellular, opening the way to the development of higher life forms.

Table 7.4 Major components of the modern atmosphere of Earth.

Gas	Percentage
N_2	78.084
O_2	20.946
Ar	0.9340
CO_2	0.0397
Ne	0.001818
He	0.000524
CH_4	0.000179

Source: Data from national oceanic and atmospheric administraties (NOAA) Earth System Research Laboratory.

The full biochemical mechanism of aerobic respiration is complex, but the net result is the conversion of oxygen and hydrocarbons present in the cell as glucose, fat or protein into carbon dioxide and water. The simplified reaction for glucose, $C_6H_{12}O_6$, is:

$$C_6H_{12}O_6 + 6O_2 \rightarrow 6CO_2 + 6H_2O \qquad (7.7)$$

Aerobic respiration produces energy more efficiently than anaerobic respiration, and therefore represents an evolutionary improvement. The development of aerobic respiration paved the way for the evolution of much larger organisms than had existed previously, eventually leading to population of the Earth with the vast diversity of plant and animal life we find on our planet today.

Our present-day atmosphere is almost entirely controlled by biological processes, and consists primarily of O_2 and N_2 in contact with liquid water in proportions that would be very far from equilibrium in the absence of life. The major components of the atmosphere are listed in Table 7.4.

7.9 Fossilisation and fossil fuels

As we have seen in the previous section, the evolution of life has completely transformed the composition of the Earth's atmosphere over the past 2 billion years or so. Apart from the obvious signs of life at the Earth's surface, living organisms have also left their mark in a more subtle way within the upper layers of the Earth's crust.

In most cases, when a plant or animal dies, the remains are either consumed by other organisms or undergo rapid aerobic bacterial decomposition into much simpler compounds that are returned to the biosphere and recycled. In some cases, for example if the remains have settled to the bottom of a sea or lake bed, decomposition occurs anaerobically over much longer timescales. The organic matter becomes buried under layers of sediment, and the anaerobic chemical reactions occurring over geological timescales at the resulting elevated temperatures and pressures yield a variety of hydrocarbons as products. These hydrocarbons become trapped in rocks below the Earth's surface as oil, coal, and other fossil fuels. Anaerobic decay of phytoplankton and zooplankton tend to yield petroleum and natural gas, whereas plants tend to form coal and methane.

In even rarer cases, the remains of an organism become buried soon after death in sand, mud, or soil, before significant decay has occurred. Under these conditions fossilisation can occur. There are several different mechanisms for fossilisation. For example, in *permineralisation*, groundwater permeates the remains, and minerals from the groundwater precipitate out into available spaces within the organism. While the organic matter eventually decays, the mineral deposits are essentially permanent, solid, rock. This process can occur at the single cell level, producing extremely detailed fossils. Sometimes, under favourable conditions, the organism acts as a nucleus for precipitation of minerals rather than simply acting as an 'empty shell' to be filled. In this case, mineralisation is often rapid relative to the rate of decay, and extremely fine detail can be retained in the resulting fossil. Alternatively, the organic matter may decompose in such a way as to leave an organism-shaped impression or 'mold' in the rock. This may then become filled with other minerals to form a 'cast'.

Fossils can be dated by radiometric or radioactive dating. If the radioactive half life, $t_{1/2}$ for a naturally occurring isotope present in the fossil is known, the observed abundance of the isotope and its decay products allows the date at which the fossil was formed to be determined. The abundances N_p and N_d of the parent and daughter nuclides are generally determined mass spectrometrically, and can be related to the age of the fossil, t, by the following expression:

$$N_d(t) = N_d(0) + N_p(0)(1 - e^{-t/\tau}) \tag{7.8}$$

where $N_p(0)$ and $N_d(0)$ are the number of atoms of the parent and daughter isotopes at time zero (i.e. when the fossil was formed), and $\tau = t_{1/2}/\ln 2$ is the exponential decay time constant of the radioactive decay. Often, this relationship is rewritten in terms of the measured abundance of parent ions, $N_p(t)$ rather than the initial abundance $N_p(0)$.

$$N_d(t) = N_d(0) + N_p(t)(e^{t/\tau} - 1) \qquad (7.9)$$

Fossilisation is a very rare event, and it is believed that fossils discovered to date represent considerably fewer than one percent of all the species that have ever existed on Earth.[17] However, the fossil record spans well over 3 billion years of our planet's history, and even these limited snapshots of organisms that have existed in the past have provided us with fascinating insights into the evolution of life on Earth.

7.10 Other Solar systems

We have looked in some detail at the formation of our Solar System, and at the evolution of Earth from its early days as a molten ball of rock to our present-day planet, inhabited by a diverse and ever-evolving population of plant and animal life. As we have seen, formation of the Solar System around our Sun was an inevitable consequence of the events that led to the birth of the Sun from the huge, dusty, gaseous Solar nebula and the physical conditions of temperature and pressure within the nebula and the subsequent protoplanetary disk. There was nothing unique about this sequence of events, or the physical conditions within the nebula. Our Sun is one of an estimated 7×10^{22} stars in the observable Universe,[18] and we should expect that most, if not all, of these stars are orbited by one or more planets. Finding these extrasolar planets proved to be challenging, and the first planet outside of our own Solar System was not detected until 1992, when several planets of similar size to Earth were observed[19] orbiting a pulsar known as B1257+12. The presence of the planets was determined following the observation of irregularities in the pulsation period of the pulsar. Three years later, in 1995, came the first observation of a planet orbiting a main-sequence star,[20] 51 Pegasi. This planet, officially named 51 Pegasi b, but

---

[17] D. R. Prothero, *What the Fossils Say and Why it Matters* (Columbia University Press, 2007).

[18] P. G. van Dokkum and C. Conroy, A substantial population of low-mass stars in luminous elliptical galaxies, *Nature*, **468**, 940–942 (2010).

[19] A. Wolszczan and D. A. Frail, A planetary system around the millisecond pulsar PSR1257 + 12, *Nature*, **355**(6356), 145–147 (1992).

[20] M. Mayor and D. Queloz, A Jupiter-mass companion to a solar-type star, *Nature*, **378**(6555), 355–359 (1995).

also referred to unofficially as Bellerophon or Dimidium,[21] was detected via the 'wobble' it induced in its parent star. While we often think of planets as orbiting around stars, in reality a planet and its star both orbit about their centre of mass. Because the mass of the star is generally much greater than that of the planet, the centre of mass of the system usually lies very close to the star, and the orbital motion of the star is therefore much smaller than that of the planet. Very rarely, the motion of the star can be seen directly by a telescope — the star appears to 'wobble'. However, in most cases (including 51 Pegasi) the small orbital motion is detected indirectly via a periodic oscillation in the Döppler shift of spectroscopic lines associated with emission from the star (see Section 1.2).

Over the past 20 years or so, numerous other methods of 'planet hunting' have successfully detected exoplanets. These include:

(1) The transit method: A planet passing in front of its parent star blocks some of the star's light from reaching Earth; this is visible as a periodic dip in the intensity of light detected from the star, which matches the orbital period of the planet.
(2) Gravitational lensing: A star can act as a gravitational lens, focusing light from more distant stars towards Earth. A planet orbiting the 'lens' star will cause additional gravitational lensing, making the more distant stars appear brighter.
(3) Masking: Usually the light from a star is so bright that reflected light from orbiting planets cannot be seen. However, in some cases it is possible to use a mask to block the light emitted from the star, and thereby to visualise planets directly.

The rate of detection of exoplanets increased throughout the 1990s and 2000s to a few tens of planets per year. In 2009, the Kepler Space Telescope[22] was launched, whose sole mission is to detect exoplanets via the transit method; Kepler's onboard photometer constantly monitors the brightness of around 150,000 main sequence stars, and the data is analysed for periodic reductions in brightness that may indicate the presence of one or more orbiting planets. At the time of writing, in March 2016, the

---

[21] In Greek mythology, Bellerophon rode the winged horse Pegasus. Dimidium is Latin for 'half', and the name referred to the fact that the planet has at least half the mass of Jupiter.
[22] http://www.kepler.nasa.gov.

Kepler mission has identified 1041 confirmed planets, and 4706 planetary candidates (unconfirmed planets).

In total, as of March 2016, over 2000 extrasolar planets, or 'exoplanets' have been detected and catalogued, spanning the range from small terrestrial or rocky planets, to huge gas giants. Many exoplanets are orbiting their respective stars at radii within the so-called 'habitable zone', where water is expected to be present in liquid form.

In an attempt to identify potentially habitable planets out of the many thousands that are rapidly being discovered, in 2011 Schulze-Makuch and coworkers[23] developed the ESI, an index that ranges from zero, for a planet entirely unlike Earth, to one for the Earth itself:

$$\text{ESI} = \prod_{i=1}^{n} \left(1 - \left|\frac{x_i - x_{i0}}{x_i + x_{i0}}\right|\right)^{w_i/n} \tag{7.10}$$

The product in this expression is over one or more properties such as radius, density, escape velocity, temperature, etc., x_i is the value of the relevant parameter for the exoplanet and x_{i0} is the corresponding value for Earth. The various parameters are weighted by the weighting index w_i, with $w_i = 0.50, 1.07, 0.70$, and 5.58 for the parameters of mean radius, bulk density, escape velocity, and surface temperature, respectively. Numerous planets have already been identified with ESIs above 0.8, with some close to 0.9, all of which are likely to have a rocky composition similar to that of the Earth.

The various weighting indices appearing in the ESI reflect the key requirements for the development of life. Based on the history of our own planet, we can identify a relatively small number of requirements, which, if satisfied, give life at least a fighting chance of developing on a planet. These are:

(1) The planet must be terrestrial (rocky), not a gas giant.
(2) The planet must be massive enough to hold onto an atmosphere.
(3) The temperature and pressure conditions must be appropriate for water to be present in liquid form.
(4) The planet should possess a magnetosphere to shield the planet from cosmic rays and solar wind.

---

[23] D. Schulze-Makuch, A. Méndez, A. G. Fairén, P. von Paris, C. Turse, G. Boyer, A. F. Davila, M. R. Antonio D. Catling, and L. N. Irwin. A two-tiered approach to assessing the habitability of exoplanets, *Astrobiology*, **11**(10), 1041–1052 (2011).

(5) The planet should possess an ozone layer to shield the surface from high-energy UV light.

Given the enormous number of exoplanets in the Universe, those we already know of and those yet to be discovered, it seems inconceivable that Earth is the only planet for which these simple conditions are met. Even if these conditions are satisfied only for one in a trillion planets, there must be billions upon billions of planets in the observable Universe that are teeming with plant and animal life, sharing a similar evolutionary history to that of our own planet, Earth.

7.11 Further reading

This chapter has provided a very brief overview of the Earth's history. More detailed accounts may be found in many texts, for example:

(1) *Global Biogeochemical Cycles*, eds. S. S. Butcher, R. J. Charlson, G. H. Orians, G. V. Wolfe (Academic Press Ltd., 1992).
(2) M. Pidwirny, *Early History of the Earth*, in *Fundamentals of Physical Geography*, 2nd edn. (online textbook available at http://www.physicalgeography.net/fundamentals/contents.html, 2006).
(3) P. Francis, *Volcanoes, a Planetary Perspective* (Oxford University Press, 1993).
(4) E. A. Keller, *Environmental Geology, 4th edn.* (Charles E. Merril Publishing Co., 1985).
(5) B. J. Skinner and S. C. Porter, *The Dynamic Earth, An Introduction to Physical Geology*, 3rd edn. (John Wiley & Sons Inc, 1995).

7.12 Questions

7.12.1 Essay-style questions

Q7.1 Explain how the accepted model for the formation of our Solar System explains the compositions, sizes, and orbital properties of the planets surrounding our Sun.
Q7.2 Explain why we are only able to view one face of the moon from Earth.
Q7.3 Discuss the key factors that make life on Earth possible.

Q7.4 Explain why the mineral olivine adopts different polymorphic crystal structures deep in the Earth's mantle.

Q7.5 Explain how plate tectonics have shaped the topography and composition of the Earth's crust.

Q7.6 Explain why we experience twice daily ocean tides on Earth, and why the height of the tide varies on approximately a monthly cycle.

Q7.7 Discuss ways in which the presence of liquid water has led to both physical and chemical changes in the Earth's surface.

Q7.8 When a phospholipid is added to water at low concentrations, it is found that rather than dissolving uniformly in the solvent, phospholipid molecules instead migrate to the surface of the liquid, forming an ordered structure at the interface. At higher concentrations, instead of dissolving, they form one or more of a variety of micelle structures. Explain these observations in terms of the relevant molecular interactions.

Q7.9 Discuss ways in which the development of life on Earth led to chemical changes in the Earth's surface and atmosphere.

7.12.2 Problems

P7.1 *Condensation and composition of the solar nebula*
Use Table 7.1 to determine the approximate percentage composition of condensed (solid) species in a solar nebula at temperatures of
(a) 50 K
(b) 100 K
(c) 300 K

Ignore species for which melting points are not stated in the table.

P7.2 *Earth similarity index*
The ESI is a measure of the habitability of a planet, and is given by

$$\text{ESI} = \prod_{i=1}^{n} \left(1 - \left|\frac{x_i - x_{i0}}{x_i + x_{i0}}\right|\right)^{\frac{w_i}{n}}$$

where the product is over various planetary properties that can be compared for the planet of interest and Earth, usually mean radius r, bulk density ρ, escape velocity v_{esc}, and surface temperature T_{surf}. x_i

and x_{i0} are the properties of the planet and Earth, respectively, and w_i is a weighting factor for each property i. A planet with ESI >0.8 is generally considered to be potentially habitable, while planets with 0.6 < ESI < 0.8 may be habitable.

The table below gives data for the solar planets and a randomly selected few from the ~2000 extrasolar planets that have been detected to date (EU stands for 'Earth Units'). Weighting factors w_i are given below each column header. Calculate the ESIs.

Planet	r (EU) $w_r = 0.57$	ρ (EU) $w_\rho = 1.07$	v_{esc} (EU) $w_v = 0.70$	T_{surf} (K) $w_T = 5.58$
Earth	1.00	1.00	1.00	288
Mercury	0.38	0.98	0.38	440
Venus	0.95	0.95	0.93	730
Mars	0.53	0.71	0.45	227
Jupiter	10.97	0.24	5.38	152
Saturn	9.14	0.12	3.23	134
Uranus	3.98	0.23	1.91	76
Neptune	3.87	0.3	2.11	72
GJ581g	1.36	1.22	1.51	278
GJ581c	1.6	1.36	1.87	380
61Virc	3.34	0.49	2.33	541
55cncc	5.24	0.37	3.2	535
rhoCrBb	11.17	0.24	5.44	650

P7.3 *Tidal locking of the Earth and Moon*

The time t_{lock} required in order for a satellite to achieve tidal locking to another body may be calculated (very) approximately using the following expression:

$$t_{\text{lock}} = (1.89216 \times 10^{18} \text{kg}^2 \text{s}^3 \text{m}^{-6}) \frac{a^6 R \mu}{m_s m_b^2}$$

where a is the semi-major axis of the orbit of the satellite s about the body b, R is the mean radius of the satellite, μ is the rigidity of the satellite, and m_s and m_b are the masses of the satellite and the body about which it is orbiting, respectively. Data for the Earth, Sun, and Moon are given in the table below.

	Earth	Sun	Moon
a/km	1.496×10^8 (about Sun)		
	384748 (about Moon)		
R/km	6371 km		
μ/N m^{-2}	3×10^{10}		
m/kg	5.972×10^{24}	1.989×10^{30}	7.3477×10^{22}

(a) Use the data above to estimate the time required for the Earth to achieve tidal locking to: (i) the Sun; and (ii) the Moon.

(b) Noting that the equation given is only likely to be accurate to within an order of magnitude or two, and therefore that the absolute values of t_{lock} are unlikely to be reliable (e.g. a more detailed calculation for the Earth–Moon system predicts a tidal locking time of around 50 billion years), comment on the relative values of your answers to (i) and (ii).

P7.4 *Calculating the pH of rain water*

In this problem you will work through the steps required to calculate the pH of rain water (Section 7.7) by considering the equilibrium between atmospheric and dissolved CO_2.

(a) The concentration of dissolved CO_2 may be determined using Henry's law, which relates the concentration of dissolved solute to the vapour pressure above the solution.

$$p_i = k_H c_i$$

Here, k_H is Henry's constant, equal to 29.41 dm^3 atm mol^{-1} for CO_2. Given that CO_2 makes up 0.0355% of the atmosphere on average, calculate the concentration of dissolved CO_2.

(b) In solution, CO_2 reacts to form carbonic acid, H_2CO_3. This dissociates according to the equilibrium $H_2CO_3 \rightleftharpoons H^+ + HCO_3^-$, with an equilibrium constant of 4.45×10^{-7}. Write down an expression for the equilibrium constant, and use your expression together with your result from (a) to determine the concentration (or activity) of H^+ ions, and therefore the pH of rain water.

P7.5 *Radiometric dating*

Carbon dating can be used to date the remains of living organisms that lived up to 50,000 years or so ago. A small amount of ^{14}C is formed in the atmosphere as the product of neutrons from cosmic radiation interacting with ^{14}N via the process $^{14}N + n \rightarrow\, ^{14}C + p$.

Some of this ^{14}C reacts to form CO_2, which is taken up by plants (and animals that eat plants). As a result, these living organisms have the same $^{14}C/^{12}C$ ratio as carbon in the atmosphere. When the organism dies, it no longer takes up ^{14}C, and the ratio gradually decreases as the ^{14}C undergoes β decay to ^{14}N, with a half life of 5720 years.

(a) Radiative decay is a first-order process. Determine the first-order rate constant for β decay of ^{14}C.
(b) Write down an equation relating the $^{14}C/^{12}C$ ratio $R(t)$ at time t to that at time zero, $R(0)$.
(c) Ötzi the iceman was discovered buried deep in an Austrian glacier. His $^{14}C/^{12}C$ ratio was found to be 0.5762 times that of the present-day atmosphere. How long ago did Ötzi die?

Appendix A
Rates of Chemical Reactions

The rate of a chemical reaction is defined as the rate at which reactants are used up or products are formed. To ensure that the same rate is obtained no matter which reactant or product is monitored, the rate of change is divided by the stoichiometric coefficient for the chosen species in the reaction equation. For example, the rate of the generic reaction

$$A + 3B \rightarrow 2C \qquad (A.1)$$

can be defined in any of the following ways:

$$\text{rate} = -\frac{d[A]}{dt} = -\frac{1}{3}\frac{d[B]}{dt} = \frac{1}{2}\frac{d[C]}{dt} \qquad (A.2)$$

A.1 Reactions occurring in a single step

For a chemical reaction occurring in a single step (often referred to as an *elementary process*, *elementary reaction*, or *elementary step*), the reaction rate is directly proportional to the concentrations of the reactants. The constant of proportionality is known as the *rate constant*. Expressing this relationship mathematically yields the *rate law* for the reaction. For example:

First-order reaction	$A \rightarrow B$	$d[A]/dt = -k[A]$
Second-order reaction	$A + A \rightarrow P$	$d[A]/dt = -k[A][A] = -k[A]^2$
	$A + B \rightarrow P$	$d[A]/dt = -k[A][B]$

The power to which each reactant is raised in the rate law is known as the *order* with respect to that reactant. The sum of the individual orders is the *overall order* of the reaction. Elementary steps may be *unimolecular* (and first-order), involving one reactant, *bimolecular* (and second-order), involving a collision between two reactant molecules, or in rare cases, *termolecular* (and third-order), involving a simultaneous collision between three reactants.

A rate law can be integrated to yield the concentration of the chosen reactant or product as a function of time. The resulting mathematical expressions are known, imaginatively, as *integrated rate laws*. For the rate laws in the examples above, the corresponding integrated rate laws are:

First-order reaction $A \rightarrow B$ $\ln[A] = \ln[A]_0 - kt$

Second-order reaction $A + A \rightarrow P$ $\dfrac{1}{[A]} = \dfrac{1}{[A]_0} + kt$

$A + B \rightarrow P$ $kt = \dfrac{1}{[B]_0 - [A]_0} \ln\left(\dfrac{[B]_0[A]}{[A]_0[B]}\right)$

with $[A]_0$ and $[B]_0$ the initial concentrations of reactants A and B.

Sometimes it is useful to define the *half life*, $t_{1/2}$, of a reactant, the time taken for the concentration of a reactant not initially present in excess to fall to half of its initial concentration. The half life can be found simply by substituting $[A] = [A]_0/2$ and $t = t_{1/2}$ into the appropriate integrated rate law and solving for $t_{1/2}$. For example:

First-order reaction $t_{1/2} = \dfrac{\ln 2}{k}$

Second-order reaction $t_{1/2} = \dfrac{1}{k[A]_0}$

An elementary chemical process can be thought of as a transition between two atomic or molecular states separated by a potential energy barrier. The rate constant for the process is determined by the barrier height or *activation energy*, E_a, and the energy (or temperature, T) of the reacting molecules, via the Arrhenius equation.

$$k(T) = A e^{-E_a/RT} \qquad (A.3)$$

where the constant A is known as the pre-exponential factor, and R is the gas constant. The origin of the Arrhenius equation can be explained using a number of different models, which differ primarily in their treatment of the pre-exponential factor, and vary drastically in their predictive power. In the simplest approach, known as simple collision theory, the pre-exponential factor for a bimolecular process is simply the collision frequency. More sophisticated models, such as transition state (TS) theory, take the internal structure and degrees of freedom of the reactants into account, and are considerably more successful at predicting pre-exponential factors and rate constants in agreement with values determined from experimental measurements.

A point worth noting in the context of the Arrhenius equation is that barrierless reactions have rate constants that are not temperature dependent. This is treated in some detail in Chapter 5. Most of the reactions occurring in interstellar gas clouds fall into this category, for the simple reason that reactions with significant activation barriers occur at much too slow a rate to be significant at the low temperatures typically found in these environments.

A.2 Reactions occurring in multiple steps

Multistep reactions are known as *complex reactions*, even when the mechanism is fairly simple. For reactions with multistep mechanisms, a rate law can be written for each chemical species present (reactant, product, or intermediate) to describe the overall rate of change in its concentration. These rate laws constitute a coupled system of differential equations, which can be solved to determine the concentration of each species as a function of time, and therefore the overall rate law for the system. Often the equations are complicated, and must be solved numerically. In some cases the resulting rate law takes a *simple form*, rate = $k[A]^a[B]^b[C]^c$ ..., in which the rate is proportional to the reactant and/or product concentrations each raised to some power. In others the rate law has a *complex form*, which may involve fractional powers and/or ratios of concentrations or sums of concentrations. For example, the Lindemann–Hinshelwood mechanism for collisionally activated dissociation of a molecule AB is

$$AB + M \underset{k_{-1}}{\overset{k_1}{\rightleftharpoons}} AB^* + M \tag{A.4}$$

$$AB^* \overset{k_2}{\rightarrow} A + B \tag{A.5}$$

where M can be any atom or molecule, and may be another AB molecule. The reaction has the overall (complex) rate law

$$\text{rate} = \frac{k_1 k_2 [\text{AB}][\text{M}]}{k_{-1}[\text{M}] + k_2} \tag{A.6}$$

In this case, we can define a definite order of one with respect to [AB], but there is no direct proportionality between the rate and the concentration of M, so we cannot define an order with respect to [M], or an overall order.

Finding solutions to the rate equations for a complex reaction becomes much more straightforward if the system reaches a steady state, such that reaction intermediates are formed and used up at similar rates and are therefore present at constant concentration for most of the reaction. In this case the rate of change of their concentrations can be approximated to zero. This is known as the *steady-state approximation*, or SSA. Applying the SSA generally transforms the complicated system of coupled differential rate equations in a system of simultaneous algebraic equations, which may be solved analytically in order to determine an overall rate law. As an example, consider the following two-step mechanism, in which the first step is reversible.

$$\text{A} + \text{B} \underset{k_{-1}}{\overset{k_1}{\rightleftharpoons}} \text{C} \overset{k_2}{\to} \text{D} \tag{A.7}$$

We can identify C as being a reactive intermediate, as it is neither a reactant nor a product. C is formed in the first step, and used up in the reverse of the first step to re-form reactants and in the second step to form the product, D. Summing these three contributions to the rate of change of C, and assuming that the total rate of change is zero (i.e. applying the SSA) gives

$$\frac{\text{d}[\text{C}]}{\text{d}t} = 0 = k_1[\text{A}][\text{B}] - k_{-1}[\text{C}] - k_2[\text{C}] \tag{A.8}$$

This may be solved to give [C] in terms of the reactant concentrations [A] and [B].

$$[\text{C}] = \frac{k_1}{k_{-1} + k_2}[\text{A}][\text{B}] \tag{A.9}$$

The overall rate of reaction is the rate of formation of the product, D, giving

$$\text{rate} = \frac{d[D]}{dt} = k_2[C] = \frac{k_1 k_2}{k_{-1} + k_2}[A][B] \tag{A.10}$$

A.3 Experimental kinetics studies

An experimental study into the kinetics of a particular reaction generally aims to determine the rate law, including the rate constant or constants and orders with respect to each reactant, and the reaction mechanism. The requirements for any experiment are that reactants are mixed and reaction initiated on a timescale that is fast relative to the rate of reaction, and that the concentrations of one or more reactants or products are tracked as a function of time. Since the rate constant is generally temperature dependent, care must be taken to ensure a constant temperature is maintained throughout the measurement. If the reaction mechanism is known, the concentration vs time data can be compared with the known rate law or integrated rate law in order to determine the rate constant, or in some cases the rate constants for the individual elementary steps. If the reaction mechanism is not known, the data can be compared with the predicted rate laws for various trial mechanisms in order to determine which mechanism is in closest agreement with experiment.

If the rate law depends on the concentrations of multiple species, it can often be simplified considerably by performing experiments under conditions in which all but one species is present in large excess. For example, if a reaction with rate law $d[A]/dt = -k[A]^a[B]^b$ is carried out with B present in large excess, the concentration of B will change very little during the experiment, and we may make the approximation that $d[A]/dt = k_{\text{eff}}[A]^a$, where $k_{\text{eff}} = k[B]^b$. The data from such an experiment may be analysed to determine the order with respect to A, as well as the effective rate constant k_{eff}, and the experiment may then be repeated with A in large excess in order to determine the order with respect to reactant B, and therefore the true bimolecular rate constant, k. This method of simplifying the rate law is known as the *isolation method*, as the dependence of the rate law on each reactant is isolated in turn. Alternatively, the dependence of the reaction rate on one reactant concentration may be determined by recording the initial rate of the reaction for various different initial concentrations of one reactant while the initial concentration of the second reactant is held constant. This is known as the *initial rates method*, and is particularly useful when secondary reactions occurring at longer times might perturb the measurement.

When the rate law depends only on the concentration of one species, either because there is only a single species reacting or because we have used the isolation method or the initial rates method to manipulate the rate law, then the rate law may be written:

$$\text{rate} = k[\text{A}]^a \qquad (A.11)$$

$$\log(\text{rate}) = \log k + a\log[\text{A}] \qquad (A.12)$$

The order a and the rate constant k may then be found from the slope and intercept, respectively, of a plot of log(rate) against log[A]. Note that since rate $= -\mathrm{d}[\text{A}]/\mathrm{d}t$, the rate data needed for this plot can be found from the slope of a plot of the measured concentration vs time.

Alternatively, the various possible integrated rate laws given above may be plotted. For example, if a reaction is first-order with respect to [A] then a plot of ln[A] vs t will be linear with a slope equal to $-k$. If a reaction is second-order with respect to [A] then a plot of $1/[\text{A}]$ vs t will be linear with a slope equal to k.

Appendix B

The Variation Principle and the Linear Variation Method

B.1 The variation principle

The variation principle states that given a system with a Hamiltonian $\hat{H}$, if ϕ is any normalised, well-behaved function that satisfies the boundary conditions of the Hamiltonian, then

$$\langle \phi | \hat{H} | \phi \rangle \geq E_0 \tag{B.1}$$

where E_0 is the true value of the lowest-energy eigenvalue of $\hat{H}$. This principle allows us to calculate an upper bound for the ground-state energy by finding the trial wavefunction ϕ for which the integral is minimised (hence the name: trial wavefunctions are varied until the optimum solution is found). Let us first verify that the variation principle is indeed correct.

We first define an integral

$$\begin{aligned} I &= \langle \phi | \hat{H} - E_0 | \phi \rangle \\ &= \langle \phi | \hat{H} | \phi \rangle - \langle \phi | E_0 | \phi \rangle \\ &= \langle \phi | \hat{H} | \phi \rangle - E_0 \langle \phi | \phi \rangle \\ &= \langle \phi | \hat{H} | \phi \rangle - E_0 \quad \text{(since } \phi \text{ is normalised)} \end{aligned} \tag{B.2}$$

If we can prove that $I \geq 0$, then we have proved the variation principle.

Let ψ_i and E_i be the true eigenfunctions and eigenvalues of $\hat{H}$, so $\hat{H}\psi_i = E_i \psi_i$. Since the eigenfunctions ψ_i form a complete basis for the space spanned by $\hat{H}$, we can expand any wavefunction ϕ in terms of the functions ψ_i (so long as ϕ satisfies the same boundary conditions as the functions ψ_i).

$$\phi = \sum_k c_k \psi_k \tag{B.3}$$

where the c_k are the expansion coefficients. Substituting this function into our integral I gives

$$I = \left\langle \sum_k c_k \psi_k \middle| \hat{H} - E_0 \middle| \sum_j c_j \psi_j \right\rangle$$

$$= \left\langle \sum_k c_k \psi_k \middle| \sum_j (\hat{H} - E_0) c_j \psi_j \right\rangle \qquad (B.4)$$

If we now use $\hat{H}\psi_j = E_j \psi_j$, we obtain

$$I = \left\langle \sum_k c_k \psi_k \middle| \sum_j c_j (E_j - E_0) \psi_j \right\rangle$$

$$= \sum_k \sum_j c_k^* c_j (E_j - E_0) \langle \psi_k | \psi_j \rangle$$

$$= \sum_k \sum_j c_k^* c_j (E_j - E_0) \delta_{jk} \qquad (B.5)$$

We now perform the sum over j, losing all terms except the $j = k$ term due to the factor δ_{jk}, to give

$$I = \sum_k c_k^* c_k (E_k - E_0)$$

$$= \sum_k |c_k|^2 (E_k - E_0) \qquad (B.6)$$

Since E_0 is the lowest eigenvalue, the quantity $E_k - E_0$ must be positive (or zero when $k = 0$), as must the quantity $|c_k|^2$. This means that all terms in the sum are non-negative, and therefore that $I \geq 0$, as required.

Note that for wavefunctions that are not normalised, the variational integral becomes:

$$\frac{\langle \phi | \hat{H} | \phi \rangle}{\langle \phi | \phi \rangle} \geq E_0 \qquad (B.7)$$

B.2 The linear variation method

A special type of variation method widely used in the study of molecules employs the so-called linear variation function, a linear combination of n linearly-independent functions $f_1, f_2, \ldots, f_n$ that satisfy the boundary conditions of the problem, i.e. $\phi = \sum_i c_i f_i$, where the c_i are the expansion coefficients. These are often chosen to be atomic orbitals. The coefficients

Appendix B: The Variation Principle and the Linear Variation Method

c_i are parameters to be determined by minimising the variational integral. In the general case, we have:

$$\langle \phi | \hat{H} | \phi \rangle = \left\langle \sum_i c_i f_i \middle| \hat{H} \middle| \sum_j c_j f_j \right\rangle$$

$$= \sum_i \sum_j c_i^* c_j \langle f_i | \hat{H} | f_j \rangle$$

$$= \sum_i \sum_j c_i^* c_j H_{ij} \qquad (B.8)$$

where $H_{ij} = \langle f_i | \hat{H} | f_j \rangle$ is the Hamiltonian matrix element, and

$$\langle \phi | \phi \rangle = \left\langle \sum_i c_i f_i \middle| \sum_j c_j f_j \right\rangle$$

$$= \sum_i \sum_j c_i^* c_j \langle f_i | f_j \rangle$$

$$= \sum_i \sum_j c_i^* c_j S_{ij} \qquad (B.9)$$

where $S_{ij} = \langle f_i | f_j \rangle$ is the overlap matrix element.

The variational energy is therefore

$$E = \frac{\sum_i \sum_j c_i^* c_j H_{ij}}{\sum_i \sum_j c_i^* c_j S_{ij}} \qquad (B.10)$$

which rearranges to give

$$E \sum_i \sum_j c_i^* c_j S_{ij} = \sum_i \sum_j c_i^* c_j H_{ij} \qquad (B.11)$$

We want to minimise the energy with respect to the linear coefficients c_i. To do this we require that $\frac{\partial E}{\partial c_i} = 0$ for all i. Differentiating both sides of Equation (B.11) gives

$$\frac{\partial E}{\partial c_k} \sum_i \sum_j c_i^* c_j S_{ij} + E \sum_i \sum_j \left[\frac{\partial c_i^*}{\partial c_k} c_j + \frac{\partial c_j}{\partial c_k} c_i^* \right] S_{ij}$$

$$= \sum_i \sum_j \left[\frac{\partial c_i^*}{\partial c_k} c_j + \frac{\partial c_j}{\partial c_k} c_i^* \right] H_{ij} \qquad (B.12)$$

Since $\frac{\partial c_i^*}{\partial c_k} = \delta_{ik}$ and $S_{ij} = S_{ji}$, $H_{ij} = H_{ji}$, we have

$$\frac{\partial E}{\partial c_k} \sum_i \sum_j c_i^* c_j S_{ij} + 2E \sum_i c_i S_{ik} = 2 \sum_i c_i H_{ik} \qquad (B.13)$$

When $\frac{\partial E}{\partial c_k} = 0$, this gives

$$\sum_i c_i(H_{ik} - ES_{ij}) = 0 \tag{B.14}$$

for all k. This equation describes a set of k *secular equations* in k unknowns. These can also be written in matrix notation, and for a non-trivial solution (i.e. $c_i \neq 0$ for all i), the determinant of the secular matrix must be equal to zero, i.e.

$$|H_{ik} - E_k S_{ik}| = 0 \tag{B.15}$$

Expanding the determinant yields an algebraic equation that can be solved to obtain the k eigenstate energies E_k. When arranged in order of increasing energy, these provide approximations to the energies of the first k states (each having an energy higher than the true energy of the state by virtue of the variation theorem). To find the energies of a larger number of states we simply use a greater number of basis functions f_i in the trial wavefunction ϕ. To obtain the approximate wavefunction for a particular state, we substitute the appropriate energy E_k into the secular equations and solve for the coefficients c_i.

NB: Using this method it is possible to find all the coefficients $c_1, \ldots, c_k$ in terms of one coefficient; normalising the wavefunction provides the absolute values for the coefficients.

Appendix C

Mass-Weighted Coordinates and the Skew Angle

The picture we presented in Chapter 5 of 'rolling a ball' or point mass across a potential energy surface (PES) worked very well in giving us a conceptual understanding of the reaction dynamics. However, when we examine this approach in a little more detail, we find that it is something of an oversimplification. We will demonstrate the problem (and how to solve it) using the triatomic A+BC system we have already considered in some detail. For this type of system it turns out that the appropriate mass for the 'rolling ball' is simply the reduced mass of the two reactants.

$$\mu = \frac{m_A m_{BC}}{m_A + m_{BC}} \tag{C.1}$$

If we choose a coordinate system (r_{AB}, r_{BC}, θ) to define the relative positions of A, B, and C, where r_{AB} and r_{BC} represent the AB and BC distances, and θ the ABC bond angle, and restrict the atoms to be colinear ($\theta = \pi$) for simplicity, then we can write the total (kinetic plus potential) energy as follows.

$$E = \frac{1}{2}\left(\mu \dot{r}_{AB}^2 + 2\frac{m_A m_C}{M}\dot{r}_{AB}\dot{r}_{BC} + \mu' \dot{r}_{BC}^2\right) + V(r_{AB}, r_{BC}, \theta = \pi) \tag{C.2}$$

where μ and μ' are the reduced masses of the reactants and products, M is the total mass, and $\dot{r} = \partial r/\partial t$.

We immediately see a problem. Using the (r_{AB}, r_{BC}) coordinate system, the kinetic energy has a cross term in $\dot{r}_{AB}\dot{r}_{BC}$, and there are two different mass factors appearing in the terms in $\dot{r}_{AB}^2$ and $\dot{r}_{BC}^2$. The kinetic energy is not of the simple form $K = \frac{1}{2}\mu v^2$, and therefore we cannot think of the motion as that of a single particle of mass μ moving across the surface.

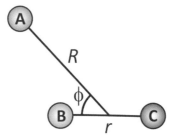

Fig. C.1   Jacobi coordinates: r is the BC distance; R is the distance from A to the centre of mass of BC; and ϕ is the angle between r and R.

However, we can rescue the situation if we define a new set of coordinates called *Jacobi coordinates*, defined in Figure C.1.

Using these new coordinates, the expression for the total energy becomes

$$E = \frac{1}{2}\mu\left(\dot{R}^2 + \alpha^{1/2}\dot{r}^2\right) + V(R, r, \phi = 0) \tag{C.3}$$

where $\alpha = \mu_{BC}/\mu$, with $\mu_{BC} = m_B m_C/m_{BC}$. The kinetic energy now does have the required simple form $K = \frac{1}{2}\mu(v_x^2 + v_y^2)$, and we find that the picture of a particle of mass μ rolling over the potential energy surface is in fact correct, *so long as* we plot the surface with the following *mass-weighted coordinates*:

$$x = R = r_{AB} + \frac{m_B}{m_{BC}} r_{BC}$$
$$y = \alpha^{1/2} r = \alpha^{1/2} r_{BC} \tag{C.4}$$

Since x and y are simply linear combinations of our original axes r_{AB} and r_{BC}, this turns out to be equivalent to plotting the potential energy surface (PES) using the (r_{AB}, r_{BC}) axis system, but with the axes at an angle β instead of at 90° to each other. The angle β is known as the skew angle, and is given by

$$\cos^2\beta = \frac{m_A m_C}{m_{AB} m_{BC}} \tag{C.5}$$

When either A or C is a light atom (i.e. light attacking or departing atom), $\cos^2\beta$ is small, β is close to 90°, and the dynamics are essentially as our original simple picture would predict. However, in other cases the skew angle can be quite small, leading to significantly different

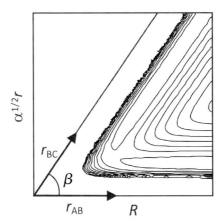

Fig. C.2 An A + BC potential energy surface plotted in mass-weighted coordinates, in which the r_{AB} and r_{BC} axes are plotted at a skew angle β. See text for details.

dynamics. Figure C.2 shows such a skewed axis system. The simple picture developed above of kinetic energy being released into translational or vibrational motion in the products becomes more complicated in such cases.

Answers to Numerical Problems

Chapter 1: Measuring the Universe

P1.1 (a) Lyman: $\Delta E = 1.635 \times 10^{-18}$ J, $\lambda = 121.5$ nm;
Balmer: $\Delta E = 3.028 \times 10^{-19}$ J, $\lambda = 656$ nm;
Paschen: $\Delta E = 1.060 \times 10^{-19}$ J, $\lambda = 1.875 \,\mu$m.
(b) Balmer, Paschen (Lyman absorbed by atmosphere).

P1.2 (a) 656.114 nm
(b) Döppler shifts are ± 0.026 nm for points on the left and right of Jupiter, 0 nm for a point in the centre.

P1.3 60 K

P1.4 $n_3/n_2 = 0.306$ (note degeneracies $g_n = n^2$ for energy levels in the H atom).

P1.5 Slope of Hubble plot gives $H = 1.37 \times 10^{-17}\,\mathrm{s}^{-1}$, yielding $t = 1/H = 2.3 \times 10^9$ years. Not very accurate!

Chapter 2: From the Big Bang to the First Atoms

P2.1 9.08×10^{-31} kg; 1.672×10^{-27} kg; 1.693×10^{-27} kg

P2.2 Proton uud, neutron udd, with one quark of each colour. Other possible combinations are uuu, ddd (Delta particles) and their antiparticles, $u\bar{u}$, $d\bar{d}$, $u\bar{d}$, $d\bar{u}$ (pions or pi mesons).

P2.3 (b) Half of the H atoms are ionized at around 9000 K.

Chapter 3: Stars and the Creation of the Higher Elements

P3.1 (a) 6.6×10^{15} m
   (b) 1.4 light years

P3.2 (a) 2.307×10^{-13} J or 1.44 MeV
   (b) 1.174×10^7 ms^{-1}, or around 4% of the speed of light.

P3.3 Looking at the emission across the visible region of the spectrum (400–700 nm), the Sun appears bright white, while Betelgeuse is much dimmer and redder.

Chapter 4: Interstellar Chemistry — Molecules in Space

P4.1 (a) 2.89×10^{35} m^{-3} s^{-1}
   (b) 1.76×10^6 m^{-3} s^{-1}
   (c) 0.0088 m^{-3} s^{-1} (32 m^{-3} hr^{-1})

P4.2 (a) (1) cosmic ray ionisation; (2) proton transfer (ion–molecule reaction); (3), (4) electron–ion dissociative recombination.
   (b) The temperature dependence agrees with the predictions of capture theory.
   (c) $$\frac{dn(H_3^+)}{dt} = k_2 n(H_2^+)n(H_2) - (k_3 + k_4)n(H_3^+)n(e^-)$$
   $$\frac{dn(H_2^+)}{dt} = k_1 n(H_2^+)n(\text{cr}) - k_2 n(H_2^+)n(H^+)$$
   (d) Equilibrium will have been reached over the lifetime of the cloud, so the steady-state approximation is valid.
   (f) 1.87×10^{-7}
   (g) H_3^+ has an extremely low proton affinity, and readily transfers a proton to other species.

P4.3 H_2 is around 10^7 times more abundant than H_2O. At a temperature of 10 K, the equations given yield a ratio of rates $k_1 n(H_2)/k_2 n(H_2)) = 1.69 \times 10^{-14}$, while at 100 K, the ratio is 5.9×10^5. Either formation pathway may therefore dominate within an interstellar cloud, depending on the local temperature.

P4.4 (b) $$\frac{dn(HCO^+)}{dt} = k_2 n(H_2)$$
   (c) (i) 1.5×10^{18} m; (ii) 160 light years

P4.5 (b) $1.134 \times 10^{-19}\,\mathrm{m}^2$
   (c) $9.038 \times 10^{-17}\,\mathrm{m}^3\,\mathrm{s}^{-1}$ or $9.038 \times 10^{-11}\,\mathrm{cm}^3\,\mathrm{s}^{-1}$
   (d) True rate constant is around 20 times higher than predicted by simple collision theory, most probably due to neglect of long-range attractive interactions between the reactants.

Chapter 5: Laboratory-Based Astrochemistry: Theory

P5.2 Microwave spectrum: CN, CO, HCl, HCN, H_2S, SO_2, NH_3, CH_3CN (molecule must have a permanent dipole).
IR spectrum: CN, CO, HCl, HCN, H_2S, SO_2, CO_2, NH_3, CH_3, CH_4, CH_3CN, C_2H_4 (molecule must have a dipole that changes during the vibration).

P5.3 For a diatomic, $I = m_1 r_1^2 + m_2 r_2^2$, where m_1 and m_2 are the atomic masses and r_1 and r_2 are the distances of the atoms from the rotation axis. Relative to the molecular centre of mass, the atomic positions are $r_1 = m_2 r/(m_1+m_2)$ and $r_2 = -m_1 r/(m_1+m_2)$, with r the bond length. Substituting these into the above equation and rearranging, recognising that $\mu = m_1 m_2/(m_1+m_2)$ yields the desired result.

P5.4 Maximising the Boltzmann distribution gives $J_{\max} = (k_B T / 2hcB)^{1/2} - 1/2$, which yields $J_{\max} = 0.84$ and 3.74 at 10 and $100\,\mathrm{K}$, respectively. Rounded down to the nearest integer to take account of quantisation, these become $J_{\max} = 0$ and 3 for $T = 10$ and $100\,\mathrm{K}$, respectively.

P5.5 H_2: $k = 570.66\,\mathrm{N\,m}^{-1}$
D_2: $k = 571.89\,\mathrm{N\,m}^{-1}$
O_2: $k = 1176.98\,\mathrm{N\,m}^{-1}$
N_2: $k = 2294.33\,\mathrm{N\,m}^{-1}$
Force constant increases in line with bond strength from single bond (H_2, D_2) to double bond (O_2) to triple bond (N_2). H_2 and D_2 have essentially identical bonding and therefore very similar force constants.

P5.6
Molecule	No. vibrational modes
H_2	1
CO_2	4
H_2O	3
NH_3	6
CH_4	9

      C_2H_2 7
      C_{60} 174

P5.8 (a) $173.5\,\mathrm{m\,s^{-1}}$
- (b) $b_{\max} = 1.294\,\mathrm{nm}$; $\sigma = 5.27 \times 10^{-10}\,\mathrm{m}^2 = 527\,\mathrm{\AA}^2$
- (c) $2.624 \times 10^{-33}\,\mathrm{kg\,m^2\,s^{-1}}$
- (d) $\Delta_r H = -382.55\,\mathrm{kJ\,mol^{-1}} = -6.353 \times 10^{-19}\,\mathrm{J}$; $E_{\mathrm{avail}} = 6.355 \times 10^{-19}\,\mathrm{J}$
- (e) $J = 3$
- (f) $E = E_{\mathrm{avail}} - E_{\mathrm{rot}} = 6.350 \times 10^{-19}\,\mathrm{J}$; $v_{\max} = 13$

P5.9 Ion–dipole: $-1.260 \times 10^{-19}\,\mathrm{J}$ or $-0.787\,\mathrm{eV}$
Dipole–dipole: $-2.542 \times 10^{-19}\,\mathrm{J}$ or $-1.59\,\mathrm{eV}$
Ion–induced dipole: $-2.094 \times 10^{-20}\,\mathrm{J}$ or $-0.131\,\mathrm{eV}$
Dipole–induced dipole: $-1.430 \times 10^{-21}\,\mathrm{J}$ or $-8.93 \times 10^{-3}\,\mathrm{eV}$
Induced dipole–induced dipole: $-2.895 \times 10^{-21}\,\mathrm{J}$ or $-1.81 \times 10^{-2}\,\mathrm{eV}$

Chapter 6: Laboratory-Based Astrochemistry: Experiment

P6.1 $2489\,\mathrm{m\,s^{-1}}$; $531\,\mathrm{m\,s^{-1}}$

P6.2 (b) (i) $2.53 \times 10^{-9}\,\mathrm{cm^{-1}}$
        (ii) $1.265 \times 10^{-5}\,\mathrm{mbar}$

P6.3 (a) $8.96\,\mathrm{MHz}$
- (b) $12.9\,\mu\mathrm{m}$
- (d) $4.0\,\mathrm{mm}$

P6.4 (b) $5.556\,\mathrm{ms}$
- (c) $1.90 \times 10^{-9}\,\mathrm{cm^3\,molecule^{-1}\,s^{-1}}$

P6.5 (b) $121.7223\,\mu\mathrm{m}$
- (c) $20.54\,\mathrm{cm^{-1}}$
- (d) Neglect of centrifugal distortion, and calculated rotational constant is $B_{v=0}$, not B_e.
- (e) $m = 19\,\mathrm{g\,mol^{-1}}$, fluorine
- (f) $65.0\,\mathrm{K}$

Chapter 7: Formation of the Solar System and the Evolution of Earth

P7.1 Only species that condense at temperatures higher than ambient will be present in condensed form at each temperature.

50 K		100 K		300 K	
Condensed species	Relative abundance	Condensed species	Relative abundance	Condensed species	Relative abundance
CO	1.00	H_2O	1.00	S	1.00
O2	0.636	CO_2	0.189	Mg	0.877
N2	0.129	NH_3	5.79×10^{-2}	Fe	0.579
H2O	1.36×10^{-2}	C_2N_2	3.84×10^{-2}	Si	0.246
CO2	2.57×10^{-3}	NO	1.79×10^{-2}		
NH3	7.86×10^{-4}	CHOOH	1.47×10^{-2}		
CH4	6.43×10^{-4}	N_2O	1.21×10^{-2}		
C2H2	5.21×10^{-4}	SO_2	8.42×10^{-3}		
NO	2.43×10^{-4}	S	3.00×10^{-4}		
CHOOH	2.00×10^{-4}	Mg	2.63×10^{-4}		
N2O	1.64×10^{-4}	Fe	1.74×10^{-4}		
SO2	1.14×10^{-4}	Si	7.37×10^{-6}		
S	4.07×10^{-6}				
Mg	3.57×10^{-6}				
Fe	2.36×10^{-6}				
Si	1.00×10^{-7}				

P7.2 Earth 1.00, Mercury 0.60, Venus 0.44, Mars 0.70, Jupiter 0.29, Saturn 0.24, Uranus 0.19, Neptune 0.18, GJ581g 0.89, GJ581c 0.70, 61Virc 0.44, 55cncc 0.38, rhoCrBb 0.25.

P7.3 (a) (i) Earth–Sun 5.44×10^9 years

(ii) Earth–Moon 1.15×10^9 years

(*Note*: These are underestimated significantly by the approximate equation given).

(b) Though the Sun is much more massive than the Moon, it is also much further away, so overall the Earth–Sun tidal forces are weaker than the Earth–Moon tidal forces.

P7.4 (a) $1.207 \times 10^{-5}\,\mathrm{mol\,dm^{-3}}$

(b) 5.63

P7.5 (a) $3.843 \times 10^{-12}\,\mathrm{s^{-1}}$

(b) 4549 years

Index

A

Ab initio calculations, 51
Absorption spectroscopy, 106
Accretion, 140–141
Accretion disk, 136
Activation energy, 176
Adiabatic capture and centrifugal sudden approximation, 88
Aerobic respiration, 163
Age of the Universe, 7
Albedo, 162
Alumina, 153
Amino acid, 160
Amino acids, interstellar space, 50
Angular momentum, 136
Anharmonic correction, 56
Anharmonicity constant, 57
Arrhenius equation, 176
Associative detachment, 37
Asteroid, 140
Asteroid belt, 141
Asthenosphere, 151–152
Asymmetric top molecule, 54, 126
Atacama Large Millimeter Array (ALMA), 140
Atmospheric modelling, 48
Atoms
    formation, 15
Aurora Australis, 149
Aurora Borealis, 149
Available energy, 62

B

Bacterial decomposition, 165
Barnard 68, 35
Basalt, 154
Basis set, 52
Beer–Lambert law, 3, 106, 122
Beta decay, 23
Big Bang, 11
Big Crunch, 12
Big Splash hypothesis, 144
Bimolecular reaction, 176
Biological molecules, interstellar medium, 49
Black body, 128, 157
Black dwarf, 24
Black hole, 25
Blue-green algae, 161
Boltzmann distribution, 3, 59, 128
Brightness temperature, 128

C

Calcium carbonate, 159
Carbon insertion reaction, 41
Carbonaceous chondrite, 156
Carbonic acid, 159
Cavity-enhanced absorption spectroscopy, 109
Cavity-enhanced spectroscopic methods, 106
Cavity-ring-down spectroscopy, 107
Cell nucleus, 163
Centrifugal barrier, 84

Centrifugal distortion coefficient, 54
Centrifugal force, 136
Chapman cycle, 163
Charge transfer, 39
Charon, 147
Chemical modelling of interstellar gas clouds, 48
Chemical weathering, 159, 162
Chu, Steve, 118
Circumstellar medium, 32
Classical trajectory, 74
Clays, 159
Closed optical loop, 119
CNO cycle, 21
CO_2 sink, 159
Cohen-Tannoudji, Claude, 118
Collision cross-section, 33, 63
  Lennard-Jones parameters, 70
Collision energy, 62
Collision frequency, 32
Collision probability, 63
Collision rate, 33, 64
Collisional dissociation, 38
Comet, 140, 142, 156
Complex rate law, 177
Complex reaction, 177
Condensation, 139
Conical intersection, 82
Conservation of angular momentum, 68, 136, 142, 146–147
Conservation of energy, 62
Continental crust, 152
Convection current, 151–152
Convergent plate boundary, 153–154
Coriolis coupling, 126
Cosmic abundance of the elements, 25
Cosmic microwave background radiation, 15
Cosmic ray ionisation, 35
Cosmological constant, 13
Coulomb crystals, 112, 118
CRESU, 111, 117
Cyanobacteria, 161
Cyclotron frequency, 114
Cyclotron motion, 113

D

Döppler cooling, 119
Döppler lineshape, 6
Döppler linewidth, 6
Döppler shift, 5, 97, 128
Dark energy, 13
Dark matter, 13
Dense molecular cloud, 31
Density of states, 111
Desorption ionization on silicon (DIOS), 111
Deuterium
  formation, 15
Diffuse interstellar bands, 50
Diffuse interstellar medium, 29
Diffuse molecular cloud, 31
Dipole operator, 59
Dispersion of chemical elements, 23
Dissociative electron attachment, 35
Dissociative recombination, 39, 43
Divergent plate boundary, 153
Dust-grain
  chemistry, 34, 37, 121
  spectrum, 37
  structure, 37

E

Earth, 140
  asthenosphere, 151
  axial tilt, 144
  core, 144, 148
  crust, 152
  elemental composition, 143
  first atmosphere, 144
  formation, 143
  layered structure, 147
  magnetic field, 148
  magnetic field reversal, 149
  mantle, 144, 149
  mantle–core boundary, 151
  mesosphere, 151
  ocean formation, 155
  primordial atmosphere, 155
  rotation period, 147
  seasons, 144

Earth similarity index, 143
Earthquake, 153
Ecliptic, 144
Effusive molecular beam, 98–99
Eigenfunction, 181
Eigenvalue, 181
Einstein coefficient, 58
Einstein, Albert, 13
Elastic collision, 60
Electron, 14
Electronic selection rules, 58
Electronic structure calculations, 51
Electrospray, 110
Elementary process, 175
Elementary reaction, 175
Elementary step, 175
Equilibrium bond length, 53
Erosion, 155, 158
Ethylene glycol, 125
Eukaryotic organism, 163
Excitation function, 65
Exoplanet detection, 143
  'wobble' method, 143
  Döppler shift, 143
  Gravitational lensing, 143
  Pulsars, 143
  transit method, 143
Expanding Universe, 11
Extrasolar planet, 143

F

Fabry–Perot cavity, 102
Fayalite, 150
Feldspar, 159
First particles, 13
Flowing afterglow, 111, 116, 120
Force constant, 54, 56
Forsterite, 150
Fossil fuels
  formation, 165
Fossilisation, 165
Fourier transform, 115
Fourier transform microwave spectroscopy, 102
Fullerenes, 51

G

Galaxy formation, 20
Geodynamo, 149
Giant gas cloud, 19
Giant impact hypothesis, 144
Giant molecular cloud, 30
Gluon, 14
Granite, 154
Gravitational collapse, 19, 135–136
Greenhouse gas, 157

H

H_2 formation on dust grains, 124
H_3^+ ion, 40
Habitable zone, 168
Half life, 165, 176
Hamiltonian, 181
Hamiltonian operator, 52
Hard-sphere limit, 67
Harmonic oscillator, 55
Helium, 138, 141
  formation, 15
Helium burning, 21
High-voltage discharge, 100
HL Tauri, 140
Hubble constant, 7
Hubble space telescope
  ultra-deep-field image, 20
Hydrogen, 138, 141
Hydrogen atom abstraction reaction, 40
Hydrogen burning, 21
Hydrologic cycle, 157

I

Ice age, 161
Ice giants, 141
Image charge, 114
Impact parameter, 62, 68
Inelastic collision, 60
Inflationary epoch, 11
Infrared spectroscopy, 35, 55, 121
Initial rates method, 180
Integrated cavity output spectroscopy, 109

Integrated rate law, 176
Interaction potential, 69
Intermolecular forces, 71
Interstellar gas cloud, 135
Interstellar ice, 121
Interstellar medium, 29
　collision rate, 33
　density, 29
Inversion symmetry, 58
Ion cyclotron resonance mass spectrometry, 111–112
Ion trap, 112
Ion–molecule reaction, 32, 34
Iron, 22, 153
Island arcs, formation, 154
Isolation method, 179

J
Jupiter, 141
　rotational speed, 6

K
Kepler space telescope, 143
Kinetic model, 48
Kuiper belt, 142

L
Langevin model, 85
Large Hadron Collider, 14
Laser ablation, 110
Laser cooling, 118
Laser pump-probe experiment, 104, 112
Laser-induced acoustic desorption (LIAD), 111
Laser-induced fluorescence, 103
Late heavy bombardment, 144, 156
Lava, 152
Laval nozzle, 117
Lennard-Jones potential, 70
Limestone, 159
Lindemann–Hinshelwood mechanism, 177
Line intensities, 3
Line positions, 1

Linear combination of atomic orbitals, 52
Linear variation method, 182
Lithosphere, 152

M
Magma, 152
Magnesium, 153
Magneto-optical trap, 119
Magnetosphere, 149
Main sequence star, 23
Mantle–core boundary, 152
Marine organisms, 159
Mars, 140
Mass spectrometry, 165
Mass-to-charge ratio, 114
Matrix isolation methods, 111
Matrix-assisted laser desorption ionization (MALDI), 111
Maxwell–Boltzmann distribution, 33, 67, 99
Mean free path, 98
Mean relative velocity, 33
Mercury, 140
Mesosphere, 152
Metallicity, 22
Meteorite, 156
Micelle, 160
Microwave pulse, 102
Microwave spectroscopy, 53, 97
Microwave transition frequencies, 54
Molecular beam, 98
　velocity distribution, 99
Molecular energy levels, 51
Molecular forces, 74
Molecular synthesis, 34
　interstellar ice, 124
Molecular wavefunctions, 51
Molecules detected in the interstellar medium, 31
Moment of inertia, 53–54
Moon
　formation, 142, 144
　Giant impact hypothesis, 144

orbit, 145
  tidal locking, 145
Morse potential, 55
Mountain range, formation, 154
Multicellular organism, 163

N

Neap tide, 157
Nebula model, 136, 142
Negative ion formation, 35
Neptune, 141
Neutral reaction, 40, 120
Neutralisation processes, 42
Neutrino, 14
Neutron, 14
  formation, 14
Neutron capture, 22
Neutron star, 25
Newton's laws of motion, 74
Nobel Prize, 13, 16, 118
Non-resonant multiphoton ionisation, 105
Northern lights, 149
Nucleosynthesis, 138

O

Ocean basin, formation, 153
Ocean formation, 156
Oceanic crust, 152
Oceanic ridge, 153
Olivine, 150
Oort cloud, 142
Optical cavity, 107
Orbital angular momentum, 67, 84
  gas cloud, 20
  projection, 58
Orbital resonance, 141
Overall rate law, 177
Oxygen atmosphere, 162
Ozone layer, 162

P

Partition function, 59
Penzias, Arno, 16
Peridot, 151

Peridotite, 150
Perlmutter, Saul, 13
Permineralisation, 165
Phillips, William, 118
Photodissociation, 38
Photoionisation, 34
Photomultiplier tube, 103
Photosynthesis, 161
Planetary nebula, 24
Planetary orbit, 135
Planetesimal, 140, 143
Pluto, 142, 147
Polanyi–Wigner equation, 123
Polycyclic aromatic hydrocarbon (PAH), 41, 51
Potassium, 155
Potential energy surface, 71
  $Ar + H_2^+$, 82
  conical intersection, 82
  construction, 73
  $H + CO_2$, 82
  $H + H_2$, 80
  $H + SO_2$, 82
  linear triatomic, 75
  long-lived complex, 82
  minima, 80
  NO_2 photolysis, 81
  reaction trajectories, 77
  saddle point, 80
  second-order saddle point, 80
Potential gradient, 74
Pre-exponential factor, 176
Primordial atmosphere, 159
Prokaryotic organism, 161
Proton, 14
  formation, 14
Proton affinity, 41
Proton capture, 21
Proton–proton cycle, 21
Proton–transfer reaction, 41
Protoplanetary disk, 136, 138, 140
Protostar, 136
Pseudo-first-order rate constant, 115
Pyroxene, 150

Q

Quantum scattering calculations, 75
Quantum state populations, 104
Quark, 14
  colour charge, 14
Quark-gluon plasma, 14
Quasi-classical trajectory, 74, 89

R

Radiative association, 36
Radical-radical reactions, 32
Radiometric dating, 165
Rate constant, 65, 111, 175
  ion–molecule reaction, 115
  neutral reaction, 120
  via Coulomb crystals, 120
  via CRESU, 118
  via flowing afterglow, 116
  via ICR-MS, 115
Rate law, 175
  experimental determination, 179
Reaction cross-section, 64, 89
Reaction order, 176
Reaction rate, 175
Reactive collision, 60
Reactive intermediate, 178
Rearrangement reaction, 42
Red giant, 24
Red supergiant, 24
Reduced mass, 56
Reflection symmetry, 58
Reflection-absorption infrared
  spectroscopy (RAIRS), 122
Relative velocity, 60
Replicating molecules, 160
Resonance-enhanced multiphoton
  ionization, 104
Rest energy, 14
Riess, Adam, 13
Rift zone, 153
Rigid rotor, 126
Ring of Fire, 153
Ring-down time, 108
Ringwoodite, 150
Rotational constant, 53–54

Rotational energy levels, 53
Rotational quantum number, 54
Rotational selection rules, 54
Rotational spectroscopy, 53
Rovibrational spectroscopy, 55, 103

S

Sagittarius B2, 125
Saturn, 141
Schmidt, Brian, 13
Schrödinger equation, 51
Secular equations, 184
Sedimentation, 155, 158
Selected-ion flow tube (SIFT), 111, 117
Selene, Moon goddess, 144
Shear force, 137
Silicates, 153
Simple collision theory, 177
Simple rate law, 177
Snowball Earth, 161
Solar nebula, 136
Solar spectrum, 1
Solar wind, 149
Southern lights, 149
Spectral windows, 4
Spectrograph, 4
Spectroscopy, 1, 97
Spin quantum number, 58
Spreading centre, 153
Spring tide, 157
Stability of interstellar molecules, 33
Star
  black dwarf, 24
  black hole, 25
  lifetimes, 20
  main sequence, 23
  metallicity, 22
  neutron star, 25
  nuclear fusion, 19
  planetary nebula, 24
  Population I, 22
  Population II, 22

Population III, 20
red giant, 24
red supergiant, 24
supernova, 25
white dwarf, 24
Star formation, 19
Steady-state approximation, 178
Stellar nursery, 19
Steric factor, 64
Subduction, 153
Sun, 22
Supernova, 21, 25, 136, 143
Supersonic molecular beam, 98, 100, 103, 110
isenthalpic expansion, 101
seeding, 101
shock-wave structure, 101
terminal velocity, 100
zone of silence, 101
Symmetric top molecule, 54

T

Tectonic plate, 151–152
Temperature-programmed desorption, 123
Termolecular reaction, 176
Terrestrial planet, 140
Theia, 144
Thermal rate constant, 89
Thorium, 155
Tidal bulge, 146
Tidal forces, 145, 157
Tidal locking, 145
Tides, 157
Transform fault boundary, 153

Transition intensity, 58
Transition state theory, 177
Translucent molecular cloud, 31
Transport processes, interstellar medium, 49
Trial wavefunction, 181
Tunnelling, 40
Tunnelling splitting, 126

U

Unimolecular reaction, 176
Uranium, 155
Uranus, 141
Urey–Miller experiment, 160

V

Vacuum energy, 13
Vapour pressure, 110
Variation principle, 52, 181
Variational transition state theory, 88
Venus, 140, 155
Vibrational frequency, 56
Vibrational modes, 57
Vibrational selection rules, 57
Vibrational wavenumber, 54
Virtual state, 105
Volcanic arc, formation, 154

W

Wadsleyite, 150
Weathering, 155
White dwarf, 24
Wilson, Robert, 16
Wollastonite, 159